D0215301

Noncooperative Game Theory

Noncooperative Game Theory

An Introduction for Engineers and Computer Scientists

João P. Hespanha

PRINCETON UNIVERSITY PRESS ~ PRINCETON AND OXFORD

Published by Princeton University Press, 41 William Street,
Princeton, New Jersey 08540

In the United Kingdom: Princeton University Press, 6 Oxford Street,
Woodstock, Oxfordshire OX20 1TR

press.princeton.edu

Jacket image: Aerial photograph of the San Francisco Bay salt ponds.
Courtesy of Robert Campbell Photography

Library of Congress Cataloging-in-Publication Data

Names: Hespanha, João P.
Title: Noncooperative game theory : an introduction for engineers and computer scientists / João P. Hespanha.
Description: Princeton : Princeton University Press, [2017] | Includes bibliographical references and index.
Identifiers: LCCN 2017001998 | ISBN 9780691175218 (hardcover : alk. paper)
Subjects: LCSH: Noncooperative games (Mathematics)—Textbooks. | Game theory—Textbooks. | Cooperative games (Mathematics)—Textbooks.
Classification: LCC QA272.5 .H47 2017 | DDC 519.3—dc23
LC record available at https://lccn.loc.gov/2017001998

British Library Cataloging-in-Publication Data is available

This book has been composed in Minion and Bank Gothic using ZzTEX
by Windfall Software, Carlisle, Massachusetts

Printed on acid-free paper. ∞

Printed in the United States of America

10 9 8 7 6 5 4 3 2 1

To my parents Graça and António

and to my sister Paula

my companions in so many games . . .

CONTENTS

III NON-ZERO-SUM GAMES

9 Two-Player Non-Zero-Sum Games

10 Computation of Nash Equilibria for Bimatrix Games

11 *N*-Player Games

12 Potential Games

13 Classes of Potential Games

IV DYNAMIC GAMES

PREAMBLE

To the uninitiated, *Game Theory* conjures images of developing computer programs to solve board games like chess or card games like poker and, in fact, the tools behind this discipline can indeed be used for such purposes. However, game theory goes much beyond such functions and provides a framework for reasoning about problems in which multiple "players" must make decisions, with the understanding that the results of these decisions affect and are affected by the decisions of the other players. Board and card games are obvious examples of such problems, but game theory is applicable to much more "serious" domains.

The first question one typically asks when faced with a multiplayer problem is probably "How should I play?" Immediately followed by the question "How will my opponent play?" The way out of this fundamental chicken and egg problem faced by game theorists is that one needs to find "equilibrium" strategies that somehow simultaneously satisfy all the players. Game theory thus provides a framework to *predict* the behavior of rational players, either in board games or in economics.

Once the notion of "equilibrium" is understood, one starts to wonder whether it is possible to find such equilibria for all games. It is not necessary to go far to find trouble: equilibria do not always exist and sometimes there is more than one equilibrium. How is one then supposed to predict the behavior of rational players? And what if a single equilibrium exists, but we do not like the predicted behavior of the players? These questions lead to some of the most interesting problems in game theory: "How to design games that predictably lead to *desirable* actions by the players?" These questions are of key interest to economists, social scientists, and engineers.

Modern game theory was born in the 1930s, mostly propelled by the work of John von Neumann, and further refined by Morgenstern, Kuhn, Nash, Shapley and others. Throughout most of the 1940s and 1950s, economics was its main application, eventually leading to the 1994 Nobel prize in Economic Science awarded to John Nash, John C. Harsanyi, and Reinhard Selten for their contributions to game theory. It was not until the 1970s that it started to have a significant impact on engineering; and in the late 1980s it led to significant breakthroughs in control theory and robust filtering. Currently, game theory pervades all areas of engineering.

Problems related to the development of pricing strategies for technological products or services have always interested engineers, but the use of game theory in technology design is a more recent development that arose from the intrinsic limitations of classical optimization-based designs. In optimization, one attempts to find values for parameters that minimize suitably defined criteria (such as monetary cost, energy consumption, heat generated, etc.) However, in most engineering applications there is always some uncertainty as to how the selected parameters will affect the final objective. One can then pose the problem of how to make sure that the selection will lead to acceptable performance, even in the presence of some degree of *uncertainty*—the unforgiving player that, behind the scenes, conspires to wreck engineering designs. This question is at the heart of many games that appear in engineering applications.

In fact, game theory provides a basic mathematical framework for robust design in engineering.

A feature common to many engineering applications of game theory is that the problem does not start as a "game." In fact, the most interesting design challenge is often to construct a game that captures the essence of the problem: Who are the players? What are their goals? Will the "solution" to the game solve the original design problem? These are questions that we will encounter in this book.

Content

Lectures 1–2 introduce the basic elements of a mathematical game through a set of simple examples. Notions of player, game rules and objectives, information structure, player rationality, cooperative versus noncooperative solutions, and Nash equilibrium are introduced to provide the reader with an overview of the main issues to come. Subsequent lectures systematically return to all of these topics.

Lectures 3–8 are focused on *zero-sum games*. Starting with matrix games in lectures 3–6, we introduce the fundamental concept of saddle-point equilibrium and explore its key properties, both for pure and mixed policies. The Minimax Theorem and computational issues are also covered. The *information structure* of a game is first treated in lectures 7–8 with the introduction of (zero-sum) games in extensive form. Complex information structures lead to the distinction between two types of stochastic policies: mixed and behavioral policies. In these lectures we also introduce a general recursive method that will evolve in later lectures into Dynamic Programming.

Non-zero sum games are treated in lectures 9–13. We introduce the concept of Nash equilibrium in a general setting and discuss its numerical computation for two-player bimatrix games. Lectures 12–13 are focused exclusively on the rich class of *potential games*. In these lectures we discuss several classical potential games, with some emphasis on the design of potential games to solve distributed optimization problems.

The last set of lectures 14–18 is devoted to the solution of dynamic games. We start by reviewing Dynamic Programming for (single-player) optimization in lectures 15–16 and use it as the starting point to construct saddle-point policies for zero-sum games in lectures 17–18. We treat both discrete- and continuous-time games, with a fixed or a variable termination time.

Learning and Teaching Using This Textbook

This book was purposely designed as a textbook, and consequently, the main emphasis is on presenting material in a fashion that makes it interesting and easy for students to understand.

Attention! When a marginal note finishes with ▷ p. *xxx* more information about that topic can be found on page *xxx*.

In writing this manuscript, there was a conscious effort to reduce verbosity. This is not to say that there was no attempt to motivate the concepts or discuss their significance (to the contrary), but the amount of text was kept to a minimum. Typically, discussion, remarks, and side comments are relegated to marginal notes so that the reader can easily follow the material presented without distraction and yet enjoy the benefit of comments on the notation and terminology, or be made aware that a there is a related MATLAB® command.

Note. João Hespanha is a Professor at the Electrical and Computer Engineering Department at the University of California, Santa Barbara.

At the University of California at Santa Barbara, I teach the material in these lectures in one quarter with about 36 hours of class time. The class I teach is primarily aimed at first-year graduate students in the College of Engineering, but these notes were written so that they can also serve as the primary textbook for a senior-level undergraduate class, as most of the lectures only require familiarity with linear algebra and probabilities at an undergraduate level. Two lectures (16 and 18) also require some familiarity with differential equations, but they could be skipped or presented as optional advanced material if students do not have the appropriate prerequisites.

I have tailored the organization of the textbook to simplify the teaching and learning of the material. In particular, the sequence of the chapters emphasizes continuity, with each chapter motivated by and in logical sequence with the preceding ones. I always avoid introducing a concept in one chapter and using it again only many chapters later. It has been my experience that even if this may be economical in terms of space, it is pedagogically counterproductive. The chapters are balanced in length so that on average each can be covered in roughly 2 hours of lecture time. Not only does this greatly aid the instructor's planning, but it makes it easier for the students to review the materials taught in class.

The book includes exercises that should be solved as the reader progresses through the material. Some of these exercises clarify issues raised in the body of the text and the reader is generally pointed to such exercises in marginal notes; for example, Exercise 6.3, which is referenced in a marginal note on page 62. Other exercises are aimed at consolidating the knowledge acquired, by asking the reader to apply algorithms or approaches previously discussed; for example, Exercise 13.9 on page 167. The book includes detailed solutions for all the exercises that appear in the sections titled "Practice Exercises," but it does not include solutions to those in the sections titled "Additional Exercises."

MATLAB®

Computational tools such as the MATLAB® software environment offer a significant step forward in teaching this class because they allow students to solve numerical problems without being subject to a detailed treatment of numerical methods. By systematically annotating the theoretical developments with marginal notes that discuss the relevant commands available in MATLAB®, this textbook helps students learn to use these tools. An example of this can be found, e.g., in MATLAB® Hint 1 on page 26, which is further expanded on page 27. We also provide MATLAB® functions that implement some of the key algorithms discussed. An example of this can be found, e.g., in MATLAB® Hint 4 on page 155, which is later used, e.g., in Exercise 13.5 on page 163.

The commands discussed in the "MATLAB® Hints" assume that the reader has version R2015b of MATLAB® and the Optimization Toolbox. However, essentially all the commands used have been fairly stable for several versions, so they are likely to work with previous and subsequent versions for several years to come. Lecture 6 assumes that the reader has installed CVX, which is a MATLAB® package for Disciplined Convex Programming, distributed under the GNU General Public License 2.0 [4].

MATLAB® is a registered trademark of The MathWorks, Inc. and is used with permission. The MathWorks does not warrant the accuracy of the text or exercises in this book. This book's use or discussion of MATLAB® or related products does not constitute an endorsement or sponsorship by The MathWorks of a particular pedagogical approach or particular use of the MATLAB® software.

Web

The reader is referred to the author's website at www.ece.ucsb.edu/~hespanha for corrections, updates on MATLAB® and CVX, and other supplemental material.

Acknowledgments

Several friends and colleagues have helped me improve this manuscript through their thoughtful constructive comments and suggestions. Among these, I owe special thanks to Jason Marden, Sérgio Pequito, Farshad Pour Safaei, as well as all the students at the University of California at Santa Barbara who used early drafts of these notes and provided me with numerous comments and suggestions. I would also like to acknowledge the support of several organizations, including the National Science Foundation (NSF), the Army Research Office (ARO), the Air Force Office of Scientific Research (AFOSR), and the University of California at Santa Barbara.

PART I INTRODUCTION

LECTURE 1

Noncooperative Games

This lecture uses several examples to introduce the key principles of noncooperative game theory.

1.1 Elements of a Game

To characterize a game one needs to specify several items:

- The *players* are the agents that make decisions.
- The *rules* define the actions allowed by the players and their effects.
- The *information structure* specifies what each player knows before making each decision.

  Chess is a *full-information* game because the current state of the game is fully known to both players as they make their decisions. In contrast, Poker is a *partial-information* game.
- The *objective* specifies the goal of each player.

For a mathematical solution to a game, one further needs to make assumptions on the *player's rationality*, regarding questions such as:

- Will the players always pursue their best interests to fulfill their objectives? [YES]
- Will the players form coalitions? [NO]
- Will the players trust each other? [NO]

The answers in square brackets characterize what are usually called *noncooperative games*, and will be implicitly assumed throughout this course. This will be further discussed shortly.

Note 1 (Human players). Studying noncooperative solutions for games played by humans reveals some lack of faith in human nature, which has certainly not prevented economists (and engineers) from doing so. However, when pursuing this approach one should not be overly surprised by finding solutions of "questionable ethics." Actually, one of the greatest contributions of noncooperative game theory is that it allows one to find problematic solutions to games and often indicates how to "fix" the games so that these solutions disappear. This type of approach falls under the heading of *mechanism design*.

In many problems, one or more players are modeling decision processes not affected by human reason, in which case one can safely pursue noncooperative solutions without questioning their ethical foundation. Robust engineering designs and evolutionary biology are good examples of this. □

1.2 Cooperative vs. Noncooperative Games: Rope-Pulling

We use the rope-pulling game to discuss the motivation and implications of assuming a noncooperative framework. This game is depicted schematically in Figure 1.1.

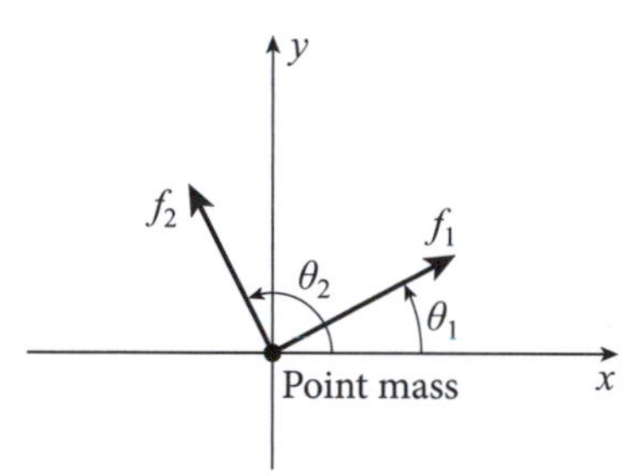

Figure 1.1 Rope-pulling game.

Rules. Two players push a point mass by exerting on it forces f_1 and f_2. Both players exert forces with the same magnitude ($|f_1| = |f_2|$), but they pull in different directions $\theta_1(t)$ and $\theta_2(t)$. The game is played for 1 second.

Note. $\theta_1(t)$ and $\theta_2(t)$ correspond to the decisions made by the players.

Initially the mass is at rest and, for simplicity, we assume unit forces and a unit mass. According to Newton's law, the point mass moves according to

Note. The equations in (1.1) effectively encode the rules of the game, as they determine how the player's decisions affect the outcome of the game.

$$\ddot{x} = \cos\theta_1(t) + \cos\theta_2(t), \qquad \dot{x}(0) = 0, \qquad x(0) = 0 \tag{1.1a}$$

$$\ddot{y} = \sin\theta_1(t) + \sin\theta_2(t), \qquad \dot{y}(0) = 0, \qquad y(0) = 0. \tag{1.1b}$$

1.2.1 Zero-Sum Rope-Pulling Game

Consider the following objective for the rope-pulling game:

Objective (zero-sum). Player P_1 wants to *maximize* $x(1)$, whereas player P_2 wants to *minimize* $x(1)$.

Notation. This is called a *zero-sum* game since players have opposite objectives. One could also imagine that P_1 wants to *maximize* $x(1)$, whereas P_2 wants to *maximize* $-x(1)$. According to this view the two objectives add up to zero.

Solution. We claim that the "optimal" solution for this game is given by

$$\mathsf{P}_1\colon \theta_1(t) = 0, \quad \forall t \in [0, 1], \mathsf{P}_2\colon \theta_2(t) = \pi, \quad \forall t \in [0, 1], \tag{1.2}$$

which results in no motion ($\ddot{x} = \ddot{y} = 0$), leading to $x(1) = y(1) = 0$ [cf. Figure 1.2(a)].

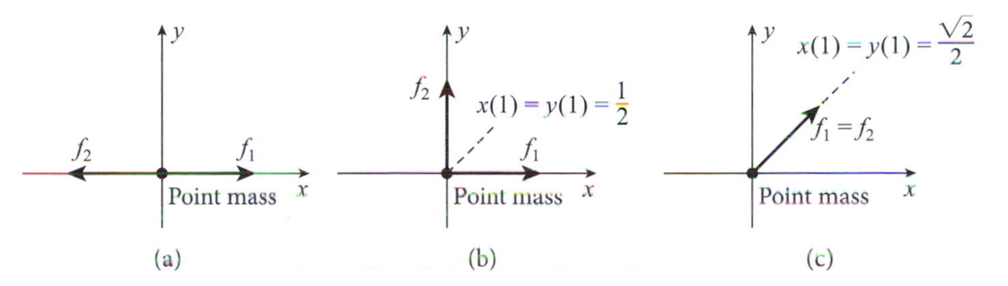

Figure 1.2 Solutions to the rope-pulling game. (a) Solution (1.2); (b) Solution (1.3); (c) Solution (1.4).

One could wonder whether it is reasonable to pull at all, given that the mass will not move. Perhaps the optimal solution is not to push at all. This is not the case for two reasons

1. Not pushing is not allowed by the rules of the game, which call for each player to exert a force of precisely one Newton, as per (1.1).
2. Even if not pulling was an option, it would be a dangerous choice for the player that decided to follow this line of action, as the other player could take advantage of the situation.

   This of course presumes that we are talking about noncooperative games for which players do not trust each other and do not form coalitions.

For now we do not justify why (1.2) is the optimal solution for the objective given above. Instead, we do this for a modified objective, for which the solution is less intuitive.

1.2.2 Non-Zero-Sum Rope-Pulling Game

Consider now a version of the game with precisely the same rules, but a modified objective:

Attention! This is no longer a zero-sum game.

Objective (non-zero-sum). Player P_1 wants to *maximize* $x(1)$, whereas player P_2 wants to *maximize* $y(1)$.

Notation. In games *a solution* is generally a set of policy, one for each player, that jointly satisfy some optimality condition.

Solution (Nash). We claim that the "optimal" solution for this game is given by

$$\mathsf{P}_1\colon \theta_1(t) = 0, \quad \forall t \in [0, 1], \qquad \mathsf{P}_2\colon \theta_2(t) = \frac{\pi}{2}, \quad \forall t \in [0, 1] \tag{1.3}$$

which leads to constant accelerations $\ddot{x} = \ddot{y} = 1$ and therefore $x(1) = y(1) = \frac{1}{2}$ [cf. Figure 1.2(b)].

This solution has two important properties:

P1.1 Suppose that player P_1 follows the course of action $\theta_1(t) = 0$ throughout the whole time period and therefore

$$\ddot{x} = 1 + \cos\theta_2(t), \quad \ddot{y} = \sin\theta_2(t), \quad \forall t \in [0, 1].$$

In this case, the best course of action for P_2 so as to maximize $y(1)$ is precisely to choose

$$\theta_2(t) = \frac{\pi}{2}, \quad \forall t \in [0, 1] \quad \Rightarrow \quad \ddot{y}(t) = 1, \quad \forall t \in [0, 1].$$

Moreover, any deviation from this will necessarily lead to a smaller value of $y(1)$. In this sense, once P_1 decides to stick to their part of the solution in (1.3), a rational P_2 must necessarily follow their policy in (1.3).

P1.2 Conversely, suppose that player P_2 follows the course of action $\theta_2(t) = \frac{\pi}{2}$ throughout the whole time period and therefore

$$\ddot{x} = \cos\theta_1(t), \quad \ddot{y} = \sin\theta_1(t) + 1, \quad \forall t \in [0, 1].$$

In this case, the best course of action for P_1 so as to maximize $x(1)$ is precisely to choose

$$\theta_1(t) = 0, \quad \forall t \in [0, 1] \quad \Rightarrow \quad \ddot{x}(t) = 1, \quad \forall t \in [0, 1].$$

Moreover, any deviation from this will necessarily lead to a smaller value of $x(1)$. Also now, once P_2 decides to stick to their part of the solution in (1.3), a rational P_1 must necessarily follow their policy in (1.3).

A pair of policies that satisfy the above properties is called a *Nash equilibrium solution.* The key feature of Nash equilibrium is that it is *stable,* in the sense that if the two players start playing at the Nash equilibrium, none of the players gains from deviating from these policies.

This solution also satisfies the following additional properties:

P1.3 Suppose that player P_1 follows the course of action $\theta_1(t) = 0$ throughout the whole time period. Then, regardless of what P_2 does, P_1 is guaranteed to achieve $x(1) \geq 0$.

Moreover, no other policy for P_1 can guarantee a larger value for $x(1)$ regardless of what P_2 does.

Note. Even if P_2 pulls against P_1, which is not very rational but possible.

P1.4 Suppose that player P_2 follows the course of action $\theta_2(t) = \frac{\pi}{2}$ throughout the whole time period. Then, regardless of what P_1 does, P_2 is guaranteed to achieve $y(1) \geq 0$.

Moreover, no other policy for P_2 can guarantee a larger value for $y(1)$ regardless of what P_1 does.

In view of this, the two policies are also called *security policies* for the corresponding player.

The solution in (1.3) is therefore "interesting" in two distinct senses: these policies form a Nash equilibrium (per P1.1–P1.2) and they are also security policies (per P1.3–P1.4).

Note. We shall see later that for zero-sum games Nash policies are always security policies (cf. Lecture 3), but this is not always the case for non-zero-sum games such as this one.

Solution (cooperative). It is also worth considering the following alternative solution

$$P_1\colon \theta_1(t) = \frac{\pi}{4}, \quad \forall t \in [0, 1], \qquad P_2\colon \theta_2(t) = \frac{\pi}{4}, \quad \forall t \in [0, 1] \tag{1.4}$$

which leads to constant accelerations $\ddot{x} = \ddot{y} = \sqrt{2}$ and therefore

$$x(1) = y(1) = \frac{\sqrt{2}}{2} > \frac{1}{2}$$

[cf. Figure 1.2(c)]. This policy is interesting because both players do strictly better than with the Nash policies in (1.3). However, this is not a Nash policy because suppose that P_1 decides to follow this course of action $\theta_1(t) = \frac{\pi}{4}$ throughout the whole time period and therefore

$$\ddot{x} = \frac{\sqrt{2}}{2} + \cos\theta_2(t), \quad \ddot{y} = \frac{\sqrt{2}}{2} + \sin\theta_2(t), \quad \forall t \in [0, 1].$$

In this case, the best course of action for P_2 to maximize $y(1)$ is to choose

$$\theta_2(t) = \frac{\pi}{2}, \quad \forall t \in [0, 1],$$

instead of the assigned policy in (1.4), because this will lead to $\ddot{y} = \frac{\sqrt{2}}{2} + 1$ and

$$y(1) = \frac{\sqrt{2} + 2}{4} > \frac{\sqrt{2}}{2}.$$

Unfortunately for P_1, this also leads to $\ddot{x} = \frac{\sqrt{2}}{2}$ and

$$x(1) = \frac{\sqrt{2}}{4} < \frac{1}{2} < \frac{\sqrt{2}}{2}.$$

In this sense, (1.4) is a dangerous choice for P_1 because a greedy P_2 will get P_1 even worse than with the Nash policy (1.3) that led to $x(1) = \frac{1}{2}$. For precisely the same reasons, (1.4) can also be a dangerous choice for P_2. In view of this, (1.4) is *not a Nash equilibrium solution*, in spite of the fact that both players can do better than with the Nash solution (1.3).

Note 2. The solution (1.4) is called *Pareto-optimal.* ▷ p. 7

Solutions such as (1.4) are the subject of *cooperative game theory*, in which one allows negotiation between players to reach a mutually beneficial solution. However, this requires faith/trust among the players. As noted above, solutions arising from cooperation are not robust with respect to cheating by one of the players.

For certain classes of games, noncooperative solutions coincide with cooperative solutions, which means that by blindingly pursuing selfish interests one actually helps other players in achieving their goals. Such games are highly desirable from a social perspective and deserve special study. It turns out that it is often possible to "reshape" the reward structure of a game to make this happen. In economics (and engineering) this is often achieved through pricing, taxation, or other incentives/deterrents and goes under the heading of *Mechanism Design.*

Note. We shall see an example in Lecture 12, where a network administrator can minimize the total interference between "selfish" wireless users by carefully charging their use of the shared medium.

Note 2 (Pareto-optimal solution). A solution like (1.4) is called *Pareto-optimal* because it is not possible to further improve the gain of one player without reducing the gain of the other. The problem of finding Pareto-optimal solutions can typically be

Note. In some cases, all Pareto-optimal solutions can also be found by solving unconstrained optimization problems. This is the case for this example, where all Pareto-optimal solutions can also be found by solving $\max_{\theta_1(t),\theta_2(t)} \beta x(1) + (1-\beta) y(1)$. The different solutions are found by picking different values for β in the interval [0, 1].

reduced to a single-criteria constrained optimization. For the non-zero-sum rope-pulling game, all Pareto-optimal solutions can be found by solving the following constrained optimization problem

$$\max_{\theta_1(t),\theta_2(t)} \{x(1) : y(1) \geq \alpha\}$$

with $\alpha \in \mathbb{R}$. Pareto-optimal solutions are generally not unique and different values of α result in different Pareto-optimal solutions. □

1.3 Robust Designs: Resistive Circuit

In many engineering applications, game theory is used as tool to solve design problems that do not start as a game. In such cases, the first step is to take the original design problem and "discover" a game theoretical formulation that leads to a desirable solution. In these games, one of the players is often the system designer and the opponent is a fictitious entity that tries to challenge the choices of the designer. Prototypical examples of this scenario are discussed in the example below and the one in Section 1.4.

Consider the resistive circuit in Figure 1.3 and suppose that our goal is to pick a resistor so that the current $i = 1/R$ is as close as possible to 1. The challenge is that when we purchase a resistor with nominal resistance equal to R_{nom}, the actual resistance may exhibit an error up to 10%, i.e.,

$$R = (1+\delta)R_{\text{nom}},$$

where δ is an unknown scalar in the interval $[-0.1, 0.1]$.

Note. Robust designs generally lead to noncooperative zero-sum games, such as this one. Cooperative solutions make no sense in robust design problems.

This a called a *robust design problem* and is often formalized as a game between the circuit designer and an unforgiving nature that does her best to foil the designer's objective:

- P_1 is the circuit designer and picks the nominal resistance R_{nom} to *minimize* the current error

$$e = \left|\frac{1}{R} - 1\right| = \left|\frac{1}{(1+\delta)R_{\text{nom}}} - 1\right|.$$

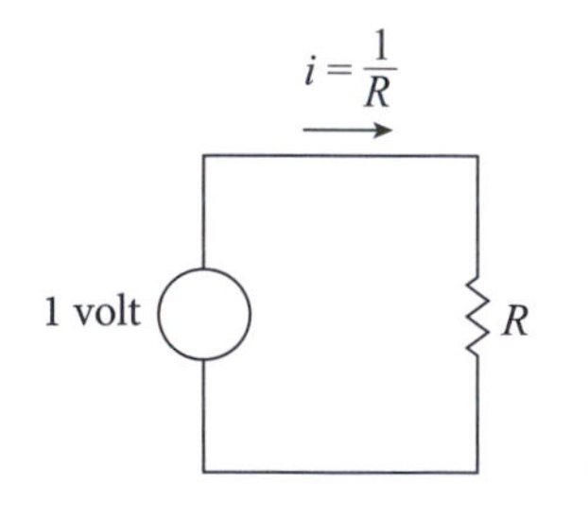

Figure 1.3 Resistive circuit game.

- P_2 is "nature" and picks the value of $\delta \in [-0.1, 0.1]$ to *maximize* the same current error e.

Solution (security). A possible solution for this game is given by

$$P_1: R_{nom} = \frac{100}{99}, \qquad P_2: \delta = 0.1 \tag{1.5}$$

which leads to a current error equal to

$$e(R_{nom}, \delta) = \left| \frac{1}{(1+\delta)R_{nom}} - 1 \right| = \left| \frac{99}{110} - 1 \right| = \left| \frac{99 - 110}{110} \right| = 0.1.$$

This solution exhibits the following properties:

P1.5 Once player P_1 picks $R_{nom} = \frac{100}{99}$, the error e will be maximized for $\delta = 0.1$ and is exactly $e = 0.1$.

P1.6 However, if player P_2 picks $\delta = 0.1$, then player P_1 can pick

$$\frac{1}{(1+\delta)R_{nom}} = 1 \;\Leftrightarrow\; R_{nom} = \frac{1}{1+\delta} = \frac{1}{1.1} = \frac{100}{110}$$

and get the error exactly equal to zero.

We thus conclude that (1.5) is *not* a safe choice for P_2 and consequently *not a Nash equilibrium*. However, (1.5) is safe for P_1 and $R_{nom} = \frac{100}{99}$ is therefore a *security policy* for the circuit designer.

Note. It turns out that, as defined, this game does not have any *Nash equilibrium*. It does however have a generalized form of Nash equilibrium that we will encounter in Lecture 4. This new form of equilibrium will allow us to "fix" P_2's policy and we shall see that P_1's choice for the resistor in (1.5) is already optimal.

1.4 Mixed Policies: Network Routing

Consider the computer network in Figure 1.4 and suppose that our goal is to send data packets from source to destination. In the typical formulation of this problem, one selects a path that minimizes the number of hops transversed by the packets. However, this formulation does not explore all possible paths and tends to create hot spots.

An alternative formulation considers two players:

- P_1 is the *router* that selects the path for the packets
- P_2 is an *attacker* that selects a link to be disabled.

The two players make their decisions independently and without knowing the choice of the other player.

Note. As in the example in Section 1.3, the second player is purely fictitious and, in this game, its role is to drive P_1 away from routing decisions that would lead to hot spots. The formulation discussed here is not unique and one can imagine other game theoretical formulations that achieve a similar goal.

Objective. Player P_1 wants to *maximize* the probability that a packet reaches its destination, whereas P_2 wants to *minimize* this probability.

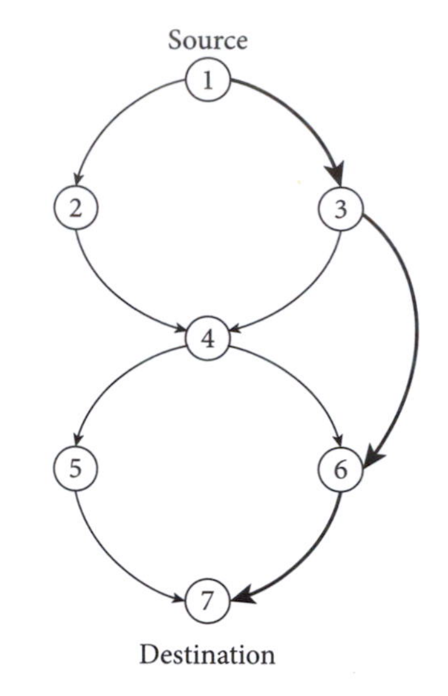

Figure 1.4 Network routing game. The 3-hop shortest path from source to destination is highlighted.

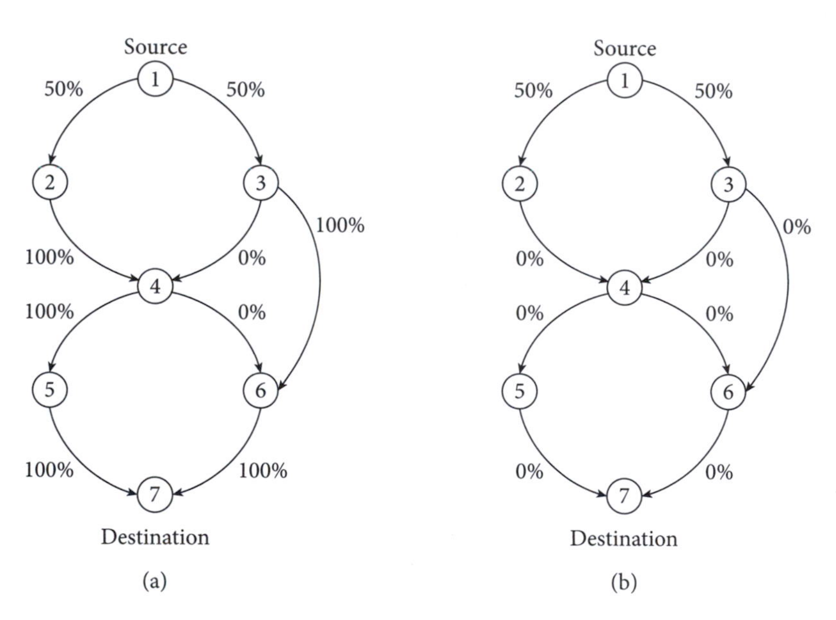

Figure 1.5 Saddle-point solution to the routing game. (a) Stochastic routing policy. The percentages by each link indicate how traffic should be distributed among the outgoing links of a node. (b) Stochastic attack policy. The percentages by each link indicate the probability by which the attacker will disable that link.

Note. See Exercise 1.1. ▷ p. 11

Solution. Figure 1.5 shows a *saddle-point solution* to this game for which 50% of the packets will reach their destination. This solution exhibits the two key properties:

P1.7 Once player P_1 picks the routing policy, P_2's attack policy is the best response from this player's perspective.

P1.8 Once player P_2 picks the routing policy, P_1's attack policy is the best response from this player's perspective.

Note. Here, "no worse" may mean "larger than" or "smaller than," depending on the player.

These are also *security policies* because each policy guarantees for that player a percentage of packet arrivals no worse than 50%. Moreover, no other policies can lead to a guaranteed better percentage of packet arrivals.

Notation. By contrast, policies that do not require randomization are called *pure policies* (cf. Lecture 4).

The policies in Figure 1.5 are *mixed policies* because they call for each player to randomize among several alternatives with "carefully" chosen probabilities. For this particular game, there are no Nash equilibrium that do not involve some form of randomization.

1.5 Nash Equilibrium

Notation. In zero-sum games, Nash equilibrium solutions are called *saddle-point solutions* (cf. Lecture 3).

The previous examples were used to illustrate the concept of *Nash equilibrium*, for which we now provide a "meta definition":

Notation. We often refer to C1.1–C1.2 as the *Nash equilibrium conditions*.

Definition 1.1 (Nash Equilibrium). Consider a game with two Players P_1, P_2. A pair of policies (π_1, π_2) is said to be a *Nash equilibrium* if the following two conditions hold:

C1.1 If P_1 uses the policy π_1, then there is no admissible policy for P_2 that does strictly better than π_2.

C1.2 If P_2 uses the policy π_2, then there is no admissible policy for P_1 that does strictly better than π_1.

Attention! Condition C1.1 does not require π_2 to be strictly better than all the other policies, just no worse. Similarly for π_1 in C1.2.

We call this a "meta definition" because it leaves open several issues that need to be resolved in the context of specific games:

Note. As we progress, we will find several notions of Nash equilibrium that fall under this general "meta definition," each corresponding to different answers to these questions.

1. What exactly is a policy?
2. What is the set of admissible policies against which π_1 and π_2 must be compared?
3. What is meant by a policy doing "strictly better" than another?

As mentioned before, the key feature of a Nash equilibrium is that it is *stable*, in the sense that if P_1 and P_2 start playing at the Nash equilibrium (π_1, π_2), none of the players gains from an unilateral deviating from these policies.

Attention! The definition of Nash equilibrium does not preclude the existence of *multiple Nash equilibria* for the same game. In fact we will find examples of that shortly (e.g., in Lecture 2).

Moreover, there are games for which there are *no Nash equilibria*.

1.6 Practice Exercise

1.1. Find other saddle-point solutions to the network routing game introduced in Section 1.4.

Solution to Exercise 1.1. Figure 1.6 shows another saddle-point solution that also satisfies the Nash equilibrium conditions C1.1–C1.2 in the (meta) Definition 1.1.

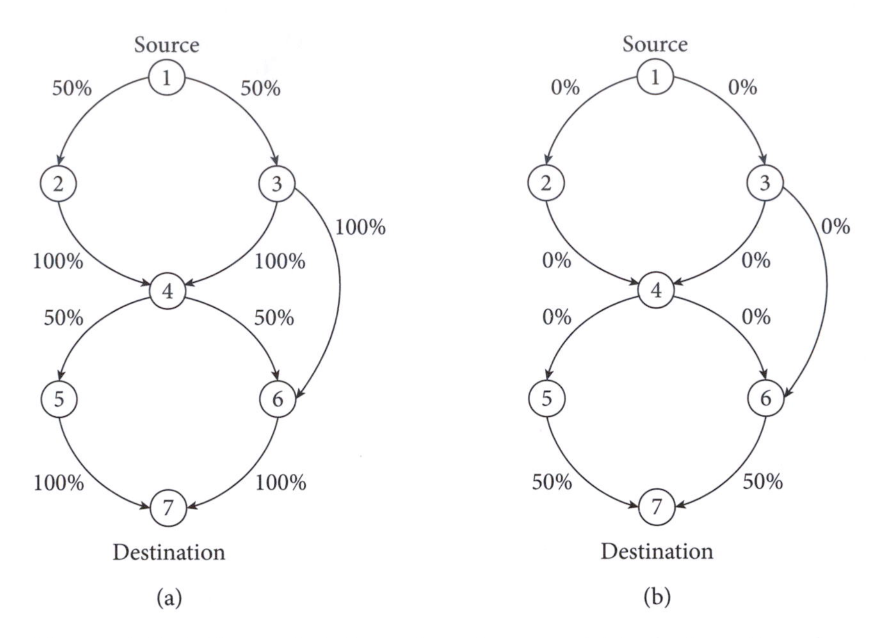

Figure 1.6 Saddle-point solution to the routing game. (a) Stochastic routing policy. The percentages by each link indicate how traffic should be distributed among the outgoing links of a node. (b) Stochastic attack policy. The percentages by each link indicate the probability by which the attacker will disable that link.

LECTURE 2

Policies

This lecture addresses the distinction between the concepts of "actions" and "policies," and several ramifications arising from this distinction.

2.1 Actions vs. Policies: Advertising Campaign

In general terms:

- An *action* is a possible move that is available to a player during a game.
- A *policy* is a decision rule that a player uses to select actions, based on available information.

Note. In game theory, it is common to use the word *strategy* as synonymous to policy.

The distinction between actions and policies is only important when players acquire information during the game that can help them in making better decisions, e.g., knowing the action of the other player. In this case, the players will *select a policy ahead of time, but they only decide on the specific actions to take, as the game evolves.*

As hinted by the terminology used in Lecture 1 the concept of Nash or saddle-point equilibrium refers to policies and not to specific actions. We will illustrate this through a well known game in marketing called the *advertising campaign.*

Rules. Consider a game played by two companies P_1, P_2 competing for the same set of clients. Each company has to decide between two possible actions: spend \$30K in an advertising campaign or keep that money as profit.

Since both companies target the same set of clients the decision of one company will affect not only its own sales, but also the sales of the other company. Table 2.1 shows the revenue for each company as a function of the actions by both players.

Note. We can see in Table 2.1(a) that when P_1 advertises, it brings clients to itself and when P_2 advertises, it steals clients from P_1. An analogous effect can be seen in Table 2.1(b).

Objective. Each player wants to maximize their profit:

$$\text{profit} = \underbrace{\text{revenue}}_{\text{from Table 2.1}} - \underbrace{\text{advertising cost}}_{0 \text{ or } \$30\text{K}}.$$

TABLE 2.1 Revenues for the Advertising Campaign Gain as a Function of the Actions of Each Company

(a) Revenue for P_1

	P_1 chooses N	P_1 chooses S
P_2 chooses N	\$169K	\$200K
P_2 chooses S	\$166K	\$185K

(b) Revenue for P_2

	P_1 chooses N	P_1 chooses S
P_2 chooses N	\$260K	\$261K
P_2 chooses S	\$300K	\$285K

In each table "S" means "spend \$30K in advertising" and "N" means "no expenditure in advertising."

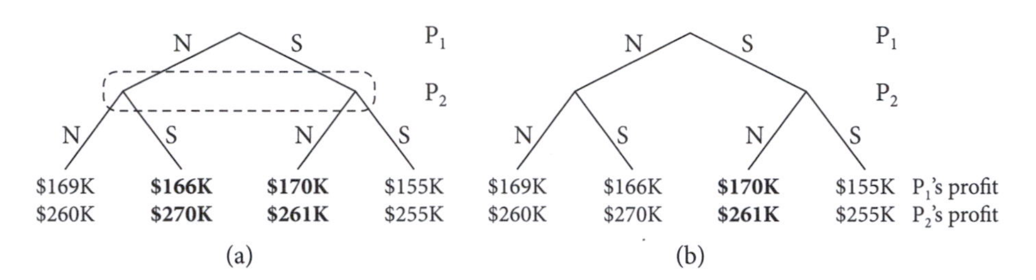

Figure 2.1 Advertising campaign game. (a) Simultaneous plays; (b) $P_1 - P_2$. The dashed box in (a) means that the corresponding player P_2 cannot distinguish between the two nodes inside the box and therefore must take the same action from all nodes inside the box. The absence of a box in (b) means that P_2 knows at which node it is at, when making a decision. The two diagrams only differ by the *information structure* of the game as implied by the dashed box. Nash equilibria solutions are denoted in boldface.

2.1.1 Simultaneous Play

We start by considering a version of the game for which both players must decide on their actions without knowing the other player's decision. In this case, the policies simply consist of deciding on a particular action, without extra information collected during the game.

Figure 2.1(a) shows a diagram of all possible choices for both players and the corresponding profits.

Notation. This type of diagram is called a game representation in *extensive form* (cf. Lecture 7).

Solution (Nash). The simultaneous play version of this game has two Nash equilibria:

$$\begin{cases} \pi_1 \equiv \mathsf{P}_1 \text{ selects N} \\ \pi_2 \equiv \mathsf{P}_2 \text{ selects S} \end{cases} \quad \text{leading to} \quad \begin{cases} \mathsf{P}_1\text{'s profit} = \$166K \\ \mathsf{P}_2\text{'s profit} = \$270K \end{cases} \tag{2.1}$$

or

$$\begin{cases} \pi_1 \equiv \mathsf{P}_1 \text{ selects S} \\ \pi_2 \equiv \mathsf{P}_2 \text{ selects N} \end{cases} \quad \text{leading to} \quad \begin{cases} \mathsf{P}_1\text{'s profit} = \$170K \\ \mathsf{P}_2\text{'s profit} = \$261K. \end{cases} \tag{2.2}$$

Note 3. Even though both policies are Nash equilibria, they are not equally favorable to both players. ▹ p. 15

Note. For games like this one with a small number of possible actions, one can find all Nash equilibria by manually checking which combination of policies satisfies the equilibria conditions. For larger games, we will need to find computational solutions (cf. Lectures 6 and 10).

It is straightforward to verify that both pairs of policies (π_1, π_2) are indeed Nash equilibria because if any one of the players unilaterally deviates from their policy, it will do strictly worse. Moreover, these are the only two Nash equilibria for this game.

Because there is no information available to the players, all policies simply consist of *unconditionally* selecting a particular action. Such policies are often called *open loop*.

Note 3 (Repeated games). Even though both policies (2.1)–(2.2) are Nash equilibria, they are not equally favorable to both players. Clearly P_1 prefers (2.2), while P_2 prefers (2.1). However, if each player plays on its favorite equilibrium (both select S), the resulting policy is not a Nash equilibrium and actually both players will profit from changing their policies. The selection of S by both players is thus said to be "unstable" and will not remain in effect for long.

What will eventually happen is the subject of an area of game theory that investigates *repeated games*. In repeated games, the players face each other by playing the same game multiple times and adjust their policies based on the results of previous games. The key question in this area is whether their policies will converge to a Nash equilibrium and, in case there are multiple equilibria, to which one of them. □

2.1.2 Alternate Play

Notation. The is often called the $P_1 - P_2$ version of the game, as opposed to the $P_2 - P_1$ version in which P_2 plays first.

Note. This type of reasoning is a very simple form of dynamic programming, which we will pursue later in a systematic fashion (cf. Lectures 7 and 17).

Consider now a version of the game in which P_1 plays first, in the sense that it selects an action before P_2, and only then P_2 plays knowing P_1's choice. Figure 2.1(b) shows a diagram with all possible choices for this game and the corresponding profits.

Solution (Nash). To find a reasonable solution for this game we place ourselves in the role of the second-playing player P_2 and hypothesize on what to do:

- If P_1 chooses N, then selecting S maximizes P_2's profit, leading to

$$\begin{cases} P_1\text{'s profit} = \$166K \\ P_2\text{'s profit} = \$270K \end{cases}$$

- If P_1 chooses S, then selecting N maximizes P_2's profit, leading to

$$\begin{cases} P_1\text{'s profit} = \$170K \\ P_2\text{'s profit} = \$261K. \end{cases} \tag{2.3}$$

Attention! The policy (2.4) is a nontrivial decision rule that a player uses to select actions, based on the available information. Such a policy is often called *closed loop*.

This motivates the following rational policy for P_2:

$$\pi_2 \equiv P_2 \text{ selects } \begin{cases} S & \text{if } P_1 \text{ selected N} \\ N & \text{if } P_1 \text{ selected S.} \end{cases} \tag{2.4}$$

If P_1 assumes that P_2 will play rationally, P_1 can guess what profit will result of their actions:

- If I choose N, then P_2 will choose S and my profit will be $166K.
- If I choose S, then P_2 will choose N and my profit will be $170K.

In view of this, the rational policy for P_1 is

$$\pi_1 \equiv P_1 \text{ selects } S,$$

leading to (2.3). One can verify that the pair of policies (π_1, π_2) so defined is a Nash equilibrium, in the sense that if any player chooses a different policy (i.e., a different decision rule for its selection of actions) while the other one keeps its policy, then the first player will do no better and potentially will do worse.

2.2 Multi-Stage Games: War of Attrition

Most of the games considered so far were *single-stage*, in the sense that each player plays only once per game. *Multi-stage games* consist of a sequence of rounds (or stages) and in each stage the players have the opportunity to take actions. In deciding the action at one stage, the players usually have available information collected from previous stages. The *war of attrition* (also known as the *chicken game*) is a typical example of a multi-stage game, which we discuss next.

Rules. This game has two players that fight repeatedly for a prize. At each stage $t \in \{0, 1, 2, \dots\}$ each player must decide to either

1. keep fighting (action "F"), for which the player incurs a cost of 1; or
2. quit (action "Q"), which means that the other player gets the prize with value $v > 1$.

   If both players quit at the same time no one gets the prize.

At each stage, the players decide on their actions simultaneously without knowing the other's decision at that stage, but knowing all the actions taken by both players at the previous stages.

Objective. Both players want to maximize their discounted profits, defined by

$$\text{P}_1\text{'s profit} = \begin{cases} \underbrace{-1-\delta-\delta^2-\cdots-\delta^{T_1-1}}_{\text{cost for playing up to time } T_1} & \text{if } \underbrace{T_1 \le T_2}_{\text{P}_1 \text{ quits first or tie}} \\ \underbrace{-1-\delta-\delta^2-\cdots-\delta^{T_2-1}}_{\text{cost for playing up to time } T_2} + \underbrace{\delta^{T_2}v}_{\text{prize collected when P}_2\text{ quits}} & \text{if } \underbrace{T_1 > T_2}_{\text{P}_2 \text{ quits first}} \end{cases}$$

$$\text{P}_2\text{'s profit} = \begin{cases} \underbrace{-1-\delta-\delta^2-\cdots-\delta^{T_2-1}}_{\text{cost for playing up to time } T_2} & \text{if } \underbrace{T_2 \le T_1}_{\text{P}_2 \text{ quits first or tie}} \\ \underbrace{-1-\delta-\delta^2-\cdots-\delta^{T_1-1}}_{\text{cost for playing up to time } T_1} + \underbrace{\delta^{T_1}v}_{\text{prize collected when P}_1\text{ quits}} & \text{if } \underbrace{T_2 > T_1}_{\text{P}_1 \text{ quits first}} \end{cases}$$

where

1. T_1 and T_2 are the times at which player P_1 and P_2, respectively quit,
2. $\delta \in (0, 1)$ is a per-step discount factor that expresses the fact that future costs/gains are less valued than current ones, i.e., paying in the future is less bad than paying now, but receiving a prize in the future is less desirable.

Note. Discount factors are well established in games in economics to take risk or inflation into account. They are also useful in engineering applications to create incentives for the players to accomplish a goal faster.

The extensive form for this game is shown in Figure 2.2.

Solution (Nash). This game has multiple Nash equilibria, some of them correspond to open-loop policies, whereas others correspond to closed-loop policies.

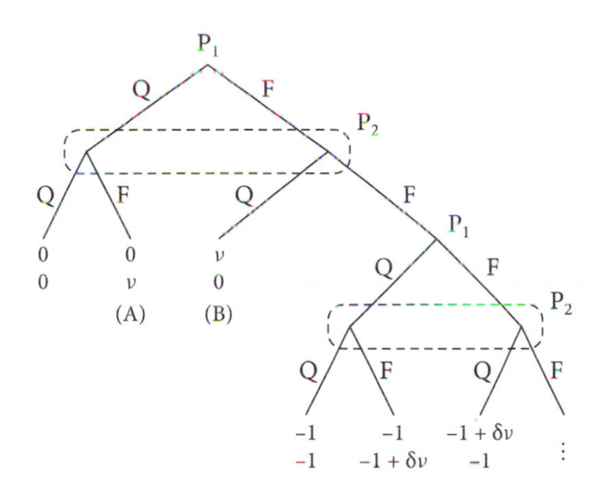

Figure 2.2 The war of attrition or the chicken game. The outcomes labeled as (A) correspond to the Nash equilibrium given by the policies (2.5), whereas the outcomes labeled as (B) correspond to the Nash equilibrium given by policies (2.6).

It is straightforward to verify that either one of the following pairs of (open-loop) policies satisfy the Nash equilibrium conditions:

$$\begin{cases} \pi_1 \equiv \mathsf{P}_1 \text{ quits as soon as possible} \\ \pi_2 \equiv \mathsf{P}_2 \text{ never quits} \end{cases} \quad \text{leading to} \quad \begin{cases} \mathsf{P}_1\text{'s profit} = 0 \\ \mathsf{P}_2\text{'s profit} = v \end{cases} \tag{2.5}$$

or

$$\begin{cases} \pi_1 \equiv \mathsf{P}_1 \text{ never quits} \\ \pi_2 \equiv \mathsf{P}_2 \text{ quits as soon as possible} \end{cases} \quad \text{leading to} \quad \begin{cases} \mathsf{P}_1\text{'s profit} = v \\ \mathsf{P}_2\text{'s profit} = 0. \end{cases} \tag{2.6}$$

Note. To verify that these are indeed Nash equilibrium we will have to wait until we study dynamic games.

Note. The policies (2.7) and (2.8) are really the same, but formulated in a different way.

It is less straightforward to verify that either one of the following pairs of policies also satisfies the Nash equilibrium conditions:

$$\begin{cases} \pi_1 \equiv \text{if } \mathsf{P}_1 \text{ did not quit before } t, \text{ quit at time } t \text{ with probability } p := \frac{1}{1+v} \\ \pi_2 \equiv \text{if } \mathsf{P}_2 \text{ did not quit before } t, \text{ quit at time } t \text{ with probability } p := \frac{1}{1+v} \end{cases} \tag{2.7}$$

or

$$\begin{cases} \pi_1 \equiv \mathsf{P}_1 \text{ quits at time } t \text{ with probability } (1-p)^t p,\ p := \frac{1}{1+v} \\ \pi_2 \equiv \mathsf{P}_2 \text{ quits at time } t \text{ with probability } (1-p)^t p,\ p := \frac{1}{1+v}. \end{cases} \tag{2.8}$$

For these pairs of policies, the profit collected by each player will be a random variable with expected values

$$\mathrm{E}[\mathsf{P}_1\text{'s profit}] = \mathrm{E}[\mathsf{P}_2\text{'s profit}] = 0.$$

These solutions raise several questions:

Note. The policies (2.7) and (2.8) are especially interesting because they correspond to a Nash equilibrium that yields the same reward to the two players, which is not possible with policies that do not involve randomization.

- Why don't both players quit at time zero, which is much simpler than (2.7) or (2.8) and yet gives the same expected profit of zero?

  This would not be a Nash equilibrium because any player that deviates from this will improve their own profit to v. The policies (2.7) and (2.8) yield the same (average) reward as quitting at time zero, but are "safer" to unilateral deviations by one of the players.

- Why don't both players never quit?

  This would also not be Nash, since they would both "collect" a profit of

$$-\sum_{t=0}^{\infty} \delta^k = -\frac{1}{1-\delta},$$

  whereas any one of them could collect a larger profit by quitting at any point.

- How to play this game?

  This is probably the wrong question to ask. A more reasonable question may be "Why should one play a game with such a pathological Nash structure?" The contribution of noncooperative game theory to this problem is precisely to reveal that the rewards structure of this game can lead to "problematic" solutions.

Note. Clearly one does not need game theory to figure this out with such a simple game, and yet . . .

2.3 Open vs. Closed-loop: Zebra in the Lake

In general terms:

- A game is *open-loop* if players must select their actions without access to any information other than what is available before the game starts.

  For such games, policies tend to simply consist of selecting a particular action, as we saw for the simultaneous play advertising campaign game in Section 2.1.1.

- A game is *closed-loop* if the players decide on their actions based on information collected after the game started.

  For such games, policies are truly decision laws, as we saw for the alternate plays advertising campaign game in Section 2.1.2.

Some games do not make sense in an open-loop fashion. This is the case of another well known game called the *zebra in the lake*, which we discuss next.

Rules. This game has two players:

1. The player P_1 is a *zebra* that swims in a circular lake with a maximum speed of v_{zebra}.
2. The player P_2 is a (hydrophobic) *lion* that runs along the perimeter of the lake with a maximum speed of $v_{\text{lion}} > v_{\text{zebra}}$.

Notation. This game falls under the class of *pursuit-evasion* games.

Objective. The two players have opposite objectives:

1. The zebra wants to get to the shore of the lake without being caught coming out of the water.
2. The lion wants to be at the precise position where the zebra leaves the lake.

Policies. In this game it is assumed that each player constantly sees the other and can react instantaneously to the current position/velocity of the other player. This game only makes sense in closed loop if the lion can see the zebra and uses this to decide where to run, because otherwise the lion has no chance of ever catching the zebra.

Note. This game is more commonly known as the "lady in the lake" and the story is that a lady is trapped in a lake and a man wants to catch her when she gets to the shore.

Solution. This game is solved in detail in [1, chapter 8.5]. However, to satisfy the reader's curiosity we provide the following information

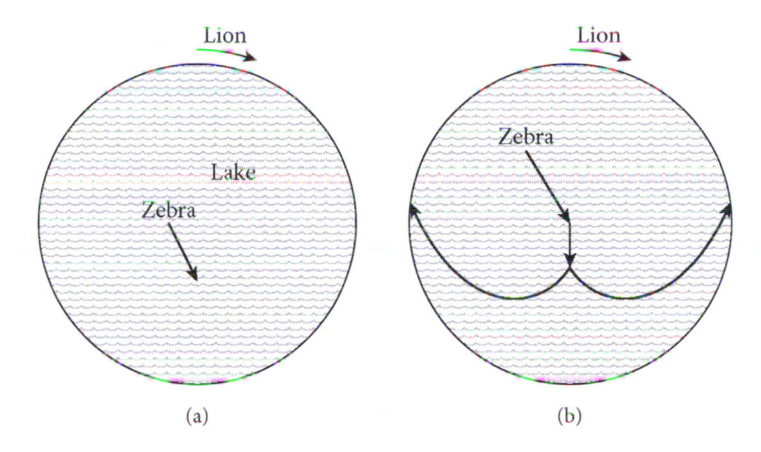

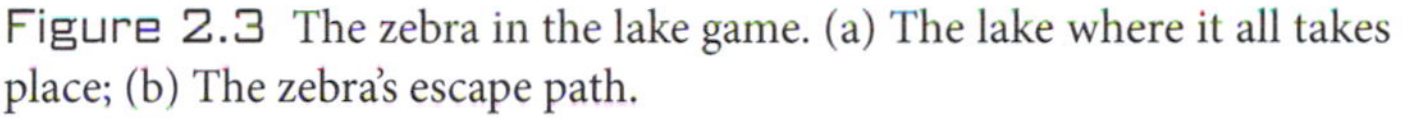
Figure 2.3 The zebra in the lake game. (a) The lake where it all takes place; (b) The zebra's escape path.

- When $v_{\text{zebra}} > .217\ v_{\text{lion}}$, the zebra can always escape.

  This is achieved by a three stage policy

  1. The zebra starts by swimming to the center of the lake.
  2. The zebra then swims towards the point in the shore opposite (furthest) from the current position of the lion. While she is close to the center of the lake she will be able to keep the (running) lion at the point in the shore furthest from her current position.
  3. As some point the zebra follows a curvy path to the shore that while not keeping the lion on the opposite side of the shore, suffices to get her to the shore safely (cf. Figure 2.3). The equation for this curve is given in [1, equation 8.38].

  The zebra's actions for the two last stages are a function of the current position of the lion.

- When $v_{\text{zebra}} \leq .217 v_{\text{lion}}$ the zebra may or may not be able to escape, depending on the initial positions of the lion and the zebra.

2.4 Practice Exercises

2.1 (Multiple Nash equilibrium). Find all Nash equilibria for the alternate plays advertising campaign game in Section 2.1.2 [cf. Figure 2.1(b)].

Solution to Exercise 2.1. The following are the only three Nash equilibria for the game:

1. One option is

$$\pi_1 \equiv \mathsf{P}_1 \text{ selects } S, \qquad \pi_2 \equiv \mathsf{P}_2 \text{ selects } \begin{cases} S & \text{if } \mathsf{P}_1 \text{ selected } N \\ N & \text{if } \mathsf{P}_1 \text{ selected } S. \end{cases}$$

This is a Nash equilibrium since if P_1 selects π_1 (i.e., S), then there is no policy for P_2 that does strictly better than π_2. Conversely, if P_2 selects π_2, then π_1 is definitely the best choice for P_1.

2. Another (simpler) option is

$$\pi_1 \equiv \mathsf{P}_1 \text{ selects } S, \qquad \pi_2 \equiv \mathsf{P}_2 \text{ selects } N.$$

This is also a Nash equilibrium since if P_1 selects π_1 (i.e., S), then there is no policy for P_2 that does strictly better than π_2 (i.e., always choosing N). Conversely, if P_2 selects π_2 (i.e., N), then π_1 is still the best choice for P_1.

3. A final option is

$$\pi_1 \equiv \mathsf{P}_1 \text{ selects } N, \qquad \pi_2 \equiv \mathsf{P}_2 \text{ selects } S.$$

This is also a Nash equilibrium since if P_1 selects π_1 (i.e., N), then there is no policy for P_2 that does strictly better than π_2 (i.e., always choosing S). Conversely, if P_2 selects π_2 (i.e., S), then π_1 is still the best choice for P_1.

2.2 (Nash equilibrium vs. order of play). Consider the game defined in extensive form in Figure 2.4. Find Nash equilibria for the following three orders of play

1. P_1 plays first ($\mathsf{P}_1 - \mathsf{P}_2$)
2. P_2 plays first ($\mathsf{P}_2 - \mathsf{P}_1$)
3. Simultaneous play.

Solution to Exercise 2.2.

1. For $\mathsf{P}_1 - \mathsf{P}_2$

$$\begin{cases} \pi_1 \equiv \mathsf{P}_1 \text{ selects } S \\ \pi_2 \equiv \mathsf{P}_2 \text{ selects } \begin{cases} S & \text{if } \mathsf{P}_1 \text{ selected N} \\ N & \text{if } \mathsf{P}_1 \text{ selected S} \end{cases} \end{cases} \quad \text{leading to} \quad \begin{cases} \mathsf{P}_1\text{'s profit} = \$170K \\ \mathsf{P}_2\text{'s profit} = \$261K \end{cases}$$

2. For $\mathsf{P}_2 - \mathsf{P}_1$

$$\begin{cases} \pi_2 \equiv \mathsf{P}_2 \text{ selects } N \\ \pi_1 \equiv \mathsf{P}_1 \text{ selects } \begin{cases} S & \text{if } \mathsf{P}_2 \text{ selected S} \\ N & \text{if } \mathsf{P}_2 \text{ selected N} \end{cases} \end{cases} \quad \text{leading to} \quad \begin{cases} \mathsf{P}_1\text{'s profit} = \$175K \\ \mathsf{P}_2\text{'s profit} = \$260K \end{cases}$$

3. For simultaneous play there is no Nash equilibrium.

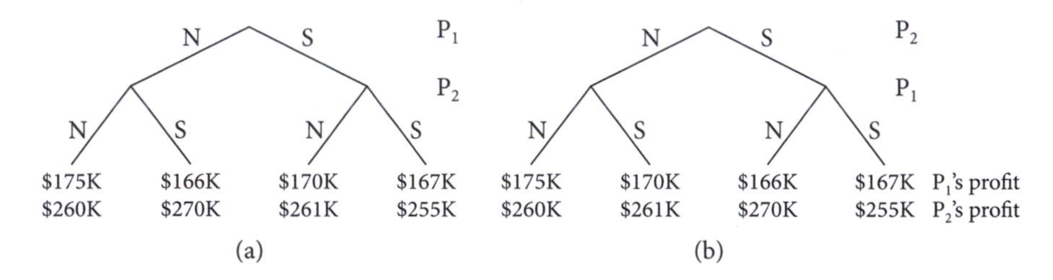

Figure 2.4 Advertising campaign game. The two diagrams correspond to the same profits, but different orders of play. (a) $\mathsf{P}_1 - \mathsf{P}_2$; (b) $\mathsf{P}_2 - \mathsf{P}_1$.

2.3 (Chicken game with alternate play). Find a Nash equilibrium policy for the chicken game with alternate play, in which P_1 plays first at each stage.

Solution to Exercise 2.3. The following policies are a Nash equilibrium:

$$\pi_1 \equiv \mathsf{P}_1 \text{ quits as soon as possible}$$

$$\pi_2 \equiv \mathsf{P}_2 \text{ never quits.}$$

PART II ZERO-SUM GAMES

LECTURE 3

Zero-Sum Matrix Games

This lecture introduces zero-sum matrix games and defines several key concepts for such games. Although simple, matrix games are the building blocks for more sophisticated games, such as the games in extensive form that we will encounter in Lecture 7.

3.1 Zero-Sum Matrix Games

Zero-sum matrix games are played by two players, each having available a finite set of actions:

- P_1 has available m actions: $\{1, 2, \ldots, m\}$
- P_2 has available n actions: $\{1, 2, \ldots, n\}$.

Notation. The set of actions available to a particular player is called the player's *action space*.

The outcome J of the game is quantified by an $m \times n$ matrix $A = [a_{ij}]$. In particular, the entry a_{ij} of the matrix provides the outcome of the game when

$$\begin{cases} \mathsf{P}_1 \text{ selects action } i \in \{1, 2, \ldots, m\} \\ \mathsf{P}_2 \text{ selects action } j \in \{1, 2, \ldots, n\}. \end{cases}$$

Note. One can imagine that P_1 selects a row of A and P_2 selects a column of A.

Objective (zero sum). The player P_1 wants to *minimize* the outcome J and the player P_2 wants to *maximize* J.

Note. This is, of course, a zero-sum game.

P_1 is often called the *minimizer*, whereas P_2 is called the *maximizer*. We consistently have the minimizer selecting rows and the maximizer selecting columns. From P_1's perspective, the outcome is called a *cost*, whereas for the perspective of P_2 the outcome is called a *reward*.

Example 3.1. The following matrix defines a zero-sum matrix game for which the minimizer has 3 actions and the maximizer has 4 actions:

$$A = \underbrace{\begin{bmatrix} 1 & 3 & 3 & -1 \\ 0 & -1 & 2 & 1 \\ -2 & 2 & 0 & 1 \end{bmatrix}}_{P_2 \text{ choices}} \Bigg\} P_1 \text{ choices}$$

□

3.2 Security Levels and Policies

A *secure* or *risk averse* playing refers to choices made by a player that are guaranteed to produce the best outcome against *any* choice made by the other player (rational or not).

For the matrix game in Example 3.1 the following are secure policies for each player

- Column 3 is a security policy for player P_2 because
  - it guarantees a reward of at least 0, and
  - no other choice can guarantee a larger reward
- Rows 2 and 3 are both security policies for player P_1 because
  - they both guarantee a cost no larger than 2, and
  - no other choice can guarantee a smaller cost.

Formally, we have the following definitions:

MATLAB® Hint 1
`min(max(A))` computes P_1's security level. ▷ p. 27

Definition 3.1 (Security policy). Consider a matrix game defined by the matrix A. The *security level* for P_1 (the minimizer) is defined by

$$\bar{V}(A) := \underbrace{\min_{i \in \{1,2,\ldots,m\}}}_{\substack{\text{minimize cost assuming} \\ \text{worst choice by } P_2}} \underbrace{\max_{j \in \{1,2,\ldots,n\}}}_{\substack{\text{worst choice by } P_2, \text{ from} \\ P_1\text{'s perspective}}} a_{ij}$$

Notation. Equation (3.1) is often written as $i^* \in \arg\min_i \max_j a_{ij}$. The use of "$\in$" instead of "$=$" emphasizes that there may be several i^* that achieve the minimum.

and the corresponding *security policy* for P_1 is any i^* that achieves the desired security level, i.e.,

$$\underbrace{\max_{j \in \{1,2,\ldots,n\}} a_{i^*j} = \bar{V}(A)}_{i^* \text{ achieves the minimum}} := \min_{i \in \{1,2,\ldots,m\}} \max_{j \in \{1,2,\ldots,n\}} a_{ij}. \tag{3.1}$$

Conversely, the *security level* for P_2 (the maximizer) is defined by

$$\underline{V}(A) := \underbrace{\max_{j \in \{1,2,\ldots,n\}}}_{\substack{\text{maximize reward} \\ \text{assuming worst choice by } P_1}} \underbrace{\min_{i \in \{1,2,\ldots,m\}}}_{\substack{\text{worst choice by } P_1, \text{ from} \\ P_2\text{'s perspective}}} a_{ij}.$$

Notation. Equation (3.2) is often written as $j^* \in \arg\max_j \min_i a_{ij}$. The use of "$\in$" instead of "$=$" emphasizes that there may be several j^* that achieve the minimum.

and the corresponding *security policy* for P_2 is any j^* that achieves the desired security level, i.e.,

$$\underbrace{\min_{i\in\{1,2,\ldots,m\}} a_{ij^*} = \underline{V}(A)}_{j^* \text{ achieves the maximum}} := \max_{j\in\{1,2,\ldots,n\}} \min_{i\in\{1,2,\ldots,m\}} a_{ij}. \tag{3.2}$$

In view of the reasoning above, we conclude that for the matrix A in Example 3.1 we have that $\underline{V}(A) = 0 \leq \bar{V}(A) = 2$.

Security levels/policies satisfy the following three simple properties:

Proposition 3.1 (Security levels/policies). For every (finite) matrix A, the following properties hold:

P3.1 Security levels are well defined and unique.

P3.2 Both players have security policies (not necessarily unique).

P3.3 The security levels always satisfy the following inequalities:

$$\underline{V}(A) := \max_{j\in\{1,2,\ldots,n\}} \min_{i\in\{1,2,\ldots,m\}} a_{ij} \leq \bar{V}(A) := \min_{i\in\{1,2,\ldots,m\}} \max_{j\in\{1,2,\ldots,n\}} a_{ij}.$$

Notation. Property P3.3 justifies the usual notation of placing a bar below or above V to denote the security levels of the maximizer and minimize, respectively. The letter V stands for "value."

Note. We shall see later games where the players have infinitely many choices, in which case we must be more careful in justifying a property similar to P3.2. ▹ p. 38

Properties P3.1 and P3.2 are trivial from the definitions. As for P3.3, it follows from the following reasoning: Let j^* be a security policy for the maximizer P_2, i.e.,

$$\underline{V}(A) = \min_i \; a_{ij^*}.$$

Since

$$a_{ij^*} \leq \max_j \; a_{ij}, \quad \forall i \in \{1, 2, \ldots, m\}$$

we conclude that

$$\underline{V}(A) = \min_i a_{ij^*} \leq \min_i \max_j a_{ij} =: \bar{V}(A),$$

which is precisely what P3.3 states. □

3.3 Computing Security Levels and Policies with MATLAB®

Note. The arguments `[],2` in the inner minimization indicate that the maximization should be performed along the second dimension of the matrix (i.e., along the columns). By default, the minimization (or maximization) is performed along the first dimension (i.e., along the rows) and therefore it is not needed to compute the security level for P_2.

MATLAB® Hint 1 (`min` and `max`). Either of the commands

```
[Vover,i]=min(max(A,[],2))
```

or

```
[Vover,i]=min(max(A'))
```

compute the pure security level `Vover` and a security policy `i` for P_1, whereas either of the commands

```
[Vunder,j]=max(min(A,[],1))
```

or

```
[Vunder,j]=max(min(A))
```

compute the pure security level `Vunder` and a security policy `j` for P_2. When more than one security policies exist, the one with the lowest index is returned. □

3.4 Security vs. Regret: Alternate Play

So far we have not yet specified the precise rules of the game, in terms of information structure (i.e., who plays first). Suppose now that the minimizer P_1 plays first ($\mathsf{P}_1 - \mathsf{P}_2$ game).

For the matrix game in Example 3.1, the optimal policy for P_2 (maximizer) is

$$\pi_2 \equiv \mathsf{P}_2 \text{ selects } \begin{cases} \text{column 2 (or 3)} & \text{if } \mathsf{P}_1 \text{ selected row 1} \quad \text{(leading to a reward of 3)} \\ \text{column 3} & \text{if } \mathsf{P}_1 \text{ selected row 2} \quad \text{(leading to a reward of 2)} \\ \text{column 2} & \text{if } \mathsf{P}_1 \text{ selected row 3} \quad \text{(leading to a reward of 2)} \end{cases} \tag{3.3}$$

in view of this, the optimal policy for P_1 (minimizer) is

$$\pi_1 \equiv \mathsf{P}_1 \text{ selects row 2 (or 3)} \quad \text{(leading to a cost of 2).} \tag{3.4}$$

If both players are rational then the outcome will be the security level for the player that plays first (P_1 in this case):

$$\bar{V}(A) = 2 \tag{3.5}$$

and no player will be surprised or regret their choice after the games end.

It would be straightforward to verify that if the maximizer P_2 plays first ($\mathsf{P}_2 - \mathsf{P}_1$ game) then the outcome will still be the security level for the player that plays first (P_2 in this case):

Note. In matrix games, the player that plays second always has the advantage: the maximizer in (3.5) and the minimizer in (3.6).

$$\underline{V}(A) = 0 \tag{3.6}$$

and again no player will regret their choice after the games end.

These conclusions generalize to *any matrix game with alternate play:* in such games, there is *no reason for rational players to ever regret their decision to play a security policy.*

3.5 Security vs. Regret: Simultaneous Plays

Suppose now that the two players for the matrix game in Example 3.1 must decide on their actions simultaneously, i.e., without knowing the others choice.

If both players use their respective *security policies* then

$$\begin{cases} \mathsf{P}_1 \text{ selects row 3} & \text{(guarantees cost} \le 2) \\ \mathsf{P}_2 \text{ selects column 3} & \text{(guarantees reward} \ge 0) \end{cases}$$

$$\text{leading to cost/reward} = 0 \in [\underline{V}(A), \bar{V}(A)].$$

However, after the game is over

- P_1 is happy since row 3 was the best response to column 3, but

- P_2 has regrets: "if I knew P_1 was going to play row 3, I would have played column 2, leading to reward $= 2 \geq 0$."

So perhaps they should have played

$$\begin{cases} P_1 \text{ selects row } 3 \\ P_2 \text{ selects column } 2 \end{cases} \quad \text{leading to cost/reward} = 2.$$

Note. The minimizer regrets the choice of row 3 in the sense that if P_1 knew that P_2 was going to play column 2, P_1 would have played row 2. However, if P_1 had indeed played row 2, then P_2 would regret (a posteriori) not having selected column 3.

But now, the minimizer regrets their choice and no further "a-posteriori" revision of the decisions would lead to a no-regret outcome.

This example leads to another important observation: as opposed to what happens in alternate play, *security policies may lead to regret in matrix games with simultaneous play*.

3.6 Saddle-Point Equilibrium

Instead of Example 3.1, consider the following matrix game:

Example 3.2. The following matrix defines a zero-sum matrix game in which both minimizer and maximizer have 2 actions:

$$A = \underbrace{\begin{bmatrix} 3 & 1 \\ -1 & 1 \end{bmatrix}}_{P_2 \text{ choices}} \Big\} P_1 \text{ choices}.$$

□

For this game

- P_1's security level is $\bar{V}(A) = 1$ and the corresponding security policy is row 2
- P_2's security level is $\underline{V}(A) = 1$ and the corresponding security policy is column 2.

Now, if both players use their respective *security policies* then

$$\begin{cases} P_1 \text{ selects row } 2 & (\text{guarantees cost} \leq 1) \\ P_2 \text{ selects column } 2 & (\text{guarantees reward} \geq 1) \end{cases}$$

$$\text{leading to cost/reward} = 1 = \underline{V}(A) = \bar{V}(A).$$

Note. Lack of regret means that one is not likely to change ones policy in subsequent games, leading to a "stable" behavior.

In this case no player regrets their choice because their policy was actually optimal against what the other player did. The same result would have been obtained in an alternate play game regardless of who plays first. This observation (lack of regret!) motivates the following definition:

Note. The qualification "pure" will be needed shortly to distinguish this for other saddle-point equilibria.

Note 4. These conditions are consistent with the Meta Definition 1.1. ▹ p. 30

Definition 3.2 (Pure saddle-point equilibrium). Consider a matrix game defined by the matrix A. A pair of policies (i^*, j^*) is called a *(pure) saddle-point equilibrium* if

$$a_{i^* j^*} \leq a_{i j^*}, \quad \forall i \tag{3.7}$$

$$a_{i^* j^*} \geq a_{i^* j}, \quad \forall j \tag{3.8}$$

and $a_{i^* j^*}$ is called the *(pure) saddle-point value*.

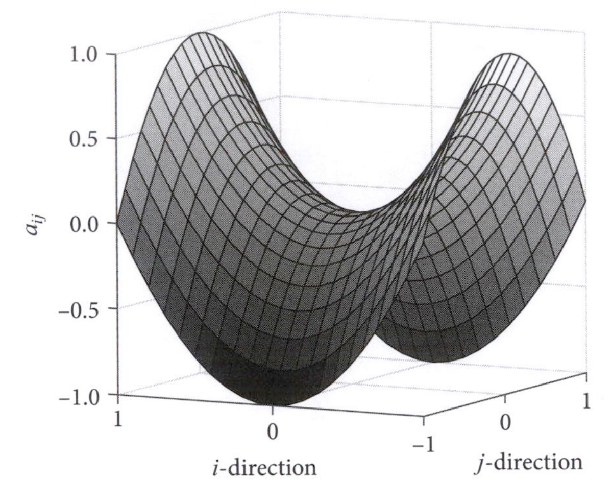

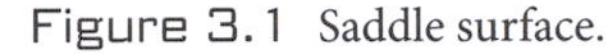
Figure 3.1 Saddle surface.

Notation. The terminology "saddle-point" arises from the fact that $a_{i^*j^*}$ increases along the "i-direction" and decreases along the "j-direction." With some imagination this corresponds to a surface that looks like a horse's saddle (cf. Figure 3.1).

Equations (3.7)–(3.8) are often re-written as

$$a_{i^*j} \leq a_{i^*j^*} \leq a_{ij^*}, \quad \forall i, j$$

and also as

$$a_{i^*j^*} = \min_i a_{ij^*} \qquad a_{i^*j^*} = \max_j a_{i^*j}. \tag{3.9}$$

Note 4 (Saddle-point equilibrium). Equation (3.7) should be interpreted as

"i^* is the best option for P_1 assuming that P_2 plays j^*,"

while (3.8) should be interpreted as

"j^* is the best option for P_2 assuming that P_1 plays i^*."

These two conditions are consistent with the Meta Definition 1.1. However, now all the hand-waving present in the meta definition has be clarified.

The two statements above could be succinctly restated as

"no player will regret her choice, if they both use these policies"

or also,

"no player will benefit from an unilateral deviation from the equilibrium." □

3.7 Saddle-Point Equilibrium vs. Security Levels

The existence of a pure saddle-point equilibrium is closely related to the security levels for the two players:

Notation. In short, the min and max commute.

Theorem 3.1 (Saddle-point equilibrium vs. security levels). A matrix game defined by A has a saddle-point equilibrium *if and only if*

$$\underline{V}(A) := \max_{j\in\{1,2,\ldots,n\}} \min_{i\in\{1,2,\ldots,m\}} a_{ij} = \min_{i\in\{1,2,\ldots,m\}} \max_{j\in\{1,2,\ldots,n\}} a_{ij} =: \bar{V}(A). \tag{3.10}$$

Notation. An important consequence of this is that all saddle-point equilibria must have exactly the same saddle-point values, which is called the *value of the game* and is denoted by $V(A)$ (without any bar). This is further discussed in Section 3.8. ▷ p. 32

Note. To prove that the two statements P and Q are equivalent, one generally starts by showing that $P \Rightarrow Q$ and then that $Q \Rightarrow P$. In *direct proofs*, to prove that $P \Rightarrow Q$ one assumes that P is true and then uses a sequence of logical steps to arrive at the conclusion that Q is also true.

In particular,

1. if (i^*, j^*) is a saddle-point equilibrium then i^* and j^* are security policies for P_1 and P_2, respectively and (3.10) is equal to the saddle-point value;
2. if (3.10) holds and i^* and j^* are security policies for P_1 and P_2, respectively then (i^*, j^*) is a saddle-point equilibrium and its value is equal to (3.10). □

To justify this theorem we start by showing that if there exists at least one saddle-point equilibrium then (3.10) must hold. To this effect, we assume that (i^*, j^*) is a saddle-point equilibrium. In this case

$$a_{i^*j^*} \underbrace{=}_{\text{by (3.9)}} \min_i a_{ij^*} \underbrace{\leq}_{\substack{\text{since } j^* \text{ is one} \\ \text{particular } j}} \max_j \min_i a_{ij} =: \underline{V}(A). \tag{3.11}$$

Similarly

$$a_{i^*j^*} \underbrace{=}_{\text{by (3.9)}} \max_j a_{i^*j} \underbrace{\geq}_{\substack{\text{since } i^* \text{ is one} \\ \text{particular } i}} \min_j \max_i a_{ij} =: \bar{V}(A). \tag{3.12}$$

Therefore

$$\bar{V}(A) \leq a_{i^*j^*} \leq \underline{V}(A).$$

But since we saw that for *every* matrix A,

$$\underline{V}(A) \leq \bar{V}(A),$$

all these inequalities are only possible if

$$\bar{V}(A) = a_{i^*j^*} = \underline{V}(A),$$

which confirms that (3.10) must always hold when a saddle point exists. In addition, since (3.11) holds with equality, we conclude that j^* must be a security policy for P_2 and since (3.12) holds with equality, i^* must be a security policy for P_1.

It remains to show that when (3.10) holds, a saddle-point equilibrium always exists. We will show that this is true and in fact the saddle-point equilibrium can be constructed by taking a security policy i^* for P_1 and a security policy j^* for P_2. Since i^* is a security policy for P_1, we have that

$$\max_j a_{i^*j} = \bar{V}(A) \quad (:= \min_j \max_i a_{ij}),$$

and since j^* is a security policy for P_2, we also have that

$$\min_i a_{ij^*} = \underline{V}(A) \quad (:= \max_j \min_i a_{ij}).$$

On the other hand, just because of what it means to be a min/max, we have that

$$\underline{V}(A) := \min_i a_{ij^*} \leq a_{i^*j^*} \leq \max_j a_{i^*j} =: \bar{V}(A).$$

When (3.10) holds, these two quantities must be equal. In particular,

$$a_{i^*j^*} = \max_j a_{i^*j} \Rightarrow a_{i^*j^*} \geq a_{i^*j}, \quad \forall j$$

$$a_{i^*j^*} = \min_i a_{ij^*} \Rightarrow a_{i^*j^*} \leq a_{ij^*}, \quad \forall i,$$

from which we conclude that (i^*, j^*) is indeed a saddle-point equilibrium.

3.8 Order Interchangeability

Suppose that a game defined by a matrix A has two distinct saddle-point equilibria:

$$(i_1^*, j_1^*) \quad \text{and} \quad (i_2^*, j_2^*).$$

In view of Theorem 3.1, both have exactly the same value $V(A) = \underline{V}(A) = \bar{V}(A)$ and

- i_1^* and i_2^* are security policies for P_1, and
- j_1^* and j_2^* are security policies for P_2.

But then again using Theorem 3.1 we conclude that the mixed pairs

$$(i_1^*, j_2^*) \quad \text{and} \quad (i_2^*, j_1^*)$$

are also saddle-point equilibria and have the same values as the original saddle points.

Proposition 3.2 (Order interchangeability). If (i_1^*, j_1^*) and (i_2^*, j_2^*) are saddle-point equilibria for a matrix game defined by A, then (i_1^*, j_2^*) and (i_2^*, j_1^*) are also saddle-point equilibria for the same game and all equilibria have exactly the same value. □

Note. This important property of zero-sum games is not shared by non-zero sum games, as we saw in Section 2.1.1 [cf. Note 3 (p. 15)]

This interchangeability property means that when one of the players finds a particular saddle-point equilibria (i_1^*, j_1^*) it is irrelevant to them whether or not the other player is playing at the same saddle-point equilibria. This is because:

1. This player will always get the same cost regardless of what saddle-point equilibrium was found by the other player.
2. Even if the other player found a different saddle-point equilibrium (i_2^*, j_2^*), there will be no regrets since the game will be played at a (third) point that is still a saddle-point.

3.9 Computational Complexity

Suppose that we want to minimize a function $f(i)$ defined over a discrete set $\{1, 2, \ldots, m\}$:

i	1	2	3	...	m
$f(i)$	$f(1)$	$f(2)$	$f(3)$	...	$f(m)$

The number of operations needed to find the minimum of f is equal to $n - 1$: one starts by comparing $f(1)$ with $f(2)$, then comparing the smallest of these with $f(3)$,

Note. In this sense, having a "good guess" for what the minimum may be, does not help from a computational point of view.

and so on. Similarly, if one suspects that a particular i^* may be a minimum of $f(i)$, one needs to perform exactly $n-1$ comparisons to verify that this is so.

Suppose now that we want to find security policies from an $m \times n$ matrix

$$A = \underbrace{\begin{bmatrix} & \vdots & \\ \cdots & a_{ij} & \cdots \\ & \vdots & \end{bmatrix}}_{n \text{ choices for } \mathsf{P}_2 \text{ (maximizer)}} \Big\} m \text{ choices for } \mathsf{P}_1 \text{ (minimizer)}.$$

To find a security policy for P_1 one needs to perform:

1. m maximizations of a function with n values, one for each possible choice of P_1 (row)
2. one minimization of the function of m values that results from the maximizations.

The total number of operations to find a security policy for P_1 is therefore equal to

$$m(n-1) + m - 1 = mn - 1,$$

which turns out to be exactly the same number of comparisons needed to find a security policy for P_2. However, suppose that one is given a candidate saddle-point equilibrium (i^*, j^*) for the game. To verify that this pair of policies is indeed a saddle-point equilibrium, it suffices to verify the saddle-point conditions (3.7)–(3.8), which only requires

$$m - 1 + n - 1 = m + n - 2$$

comparisons. If this test succeeds, we automatically obtain the two security policies, with far fewer comparisons.

Attention! The total number of comparisons is actually smaller than the number of entries of the matrix A since we actually only need to look at the row i^* and column j^* of the matrix. It is somewhat surprising that we can find security policies just by looking at a very small subset of the matrix. See Example 3.3. ▷ p. 33

Example 3.3 (Security policy for a partially known matrix game). Consider the following matrix that defines a zero-sum matrix game for which the minimizer has 4 actions and the maximizer has 6 actions:

$$A = \underbrace{\begin{bmatrix} ? & ? & ? & 2 & ? & ? \\ ? & ? & ? & 3 & ? & ? \\ -1 & -7 & -6 & 1 & -2 & -1 \\ ? & ? & ? & 1 & ? & ? \end{bmatrix}}_{\mathsf{P}_2 \text{ choices}} \Big\} \mathsf{P}_1 \text{ choices},$$

where all the '?' denote entries of the matrix that are not known. Even though we only know 9 out of the 24 entries of A, we know that the value of the game is equal to $V(A) = 1$ and that the row 3 is a security policy for P_1 and the column 4 is a security policy for P_2. □

Attention! For games, having a "good guess" for a saddle-point equilibrium, perhaps coming from some heuristics or insight into the game, can significantly reduce the computation. One should emphasize that even if the "guess" for the saddle-point equilibrium comes from heuristics that cannot be theoretically justified, one can answer precisely the question of whether or not the pair of policies is a saddle-point equilibrium and thus whether or not we have security policies, with a relatively small amount of computation.

3.10 Practice Exercise

3.1 (Pure security levels/policies). The following matrix defines a zero-sum matrix game for which both players have 4 actions:

$$A = \underbrace{\begin{bmatrix} -2 & 1 & -1 & 1 \\ 2 & 3 & -1 & 2 \\ 1 & 2 & 3 & 4 \\ -1 & 1 & 0 & 1 \end{bmatrix}}_{\mathsf{P}_2 \text{ choices}} \Bigg\} \mathsf{P}_1 \text{ choices}.$$

Compute the security levels, all security policies for both players, and all pure saddle-point equilibria (if they exist).

Solution to Exercise 3.1. For this game

$$\underline{V}(A) = 1 \qquad \text{columns 2,4 are security policies for } \mathsf{P}_2.$$

$$\bar{V}(A) = 1 \qquad \text{rows 1,4 are security policies for } \mathsf{P}_1.$$

Consequently the game has 4 pure saddle-point equilibria (1, 2), (4, 2), (1, 4), (4, 4).

3.11 Additional Exercise

3.2. Show that the pair of policies (π_1, π_2) in (3.3)–(3.4) form a Nash equilibrium in the sense of the (meta) Definition 1.1.

LECTURE 4

Mixed Policies

This lecture introduces the concept of mixed policies and shows how the notions previously defined for pure policies can be easily adapted to this more general type of policies.

4.1 Mixed Policies: Rock-Paper-Scissor

Note. You may want to check the web site of the World RPS Society at www.worldrps.com for the latest results on this year's World RPS Championship or other smaller events.

The familiar rock-paper-scissor game can be viewed as a matrix game if we make the following associations

$$\text{actions:} \begin{cases} \text{rock} & \equiv 1 \\ \text{paper} & \equiv 2 \\ \text{scissor} & \equiv 3 \end{cases} \qquad \text{outcomes:} \begin{cases} \mathsf{P}_1 \text{ wins} & \equiv -1 & \text{minimizer} \\ \mathsf{P}_2 \text{ wins} & \equiv +1 & \text{maximizer} \\ \text{draw} & \equiv 0, \end{cases}$$

which leads to the following matrix

$$A = \underbrace{\begin{bmatrix} 0 & +1 & -1 \\ -1 & 0 & +1 \\ +1 & -1 & 0 \end{bmatrix}}_{\mathsf{P}_2 \text{ choices}} \Big\} \mathsf{P}_1 \text{ choices}.$$

Note. In essence, the security levels can only "guarantee" a loss.

For this game

- P_1's security level is $\bar{V}(A) = +1$ and any row is a security policy.
- P_2's security level is $\underline{V}(A) = -1$ and any column is a security policy.

This means that we have a strict inequality $\underline{V}(A) < \bar{V}(A)$ and therefore the game has no pure saddle-point equilibria.

So far we considered only what are called *pure policies*, which consist of choices of particular actions (perhaps based on some observation). We now introduce mixed policies, which consist of choosing a probability distribution to select actions (perhaps as a function of observations).

The idea behind mixed policies is that the players select their actions randomly according to a previously selected probability distribution. Consider a game specified by an $m \times n$ matrix A, with m actions for P_1 and n actions for P_2.

- A *mixed policy for* P_1 is a set of numbers

$$\{y_1, y_2, \dots, y_m\}, \quad \sum_{i=1}^{m} y_i = 1, \quad y_i \geq 0, \quad \forall i \in \{1, 2, \dots, m\},$$

  with the understanding that y_i is the probability that P_1 uses to select the action $i \in \{1, 2, \dots, m\}$.
- A *mixed policy for* P_2 is a set of numbers

$$\{z_1, z_2, \dots, z_n\}, \quad \sum_{j=1}^{n} z_j = 1 \quad z_j \geq 0, \quad \forall j \in \{1, 2, \dots, n\},$$

  with the understanding that z_j is the probability that P_2 uses to select action $j \in \{1, 2, \dots, n\}$.

It is assumed that the random selections by both players are done statistically independently.

Due to randomness, the same pair of mixed policies will lead to different outcomes as one plays the game again and again. With mixed policies, players will then try to optimize the *expected outcome of the game*:

Notation. Since the order of summation is irrelevant we will often simply use the short hand notation $\sum_{ij}$ to denote multiple summations as in (4.1).

Notation. We use the symbol $'$ to denote matrix/vector transpose.

$$J = \sum_{i=1}^{m} \sum_{j=1}^{n} a_{ij} \text{ Prob}(\mathsf{P}_1 \text{ selects } i \text{ and } \mathsf{P}_2 \text{ selects } j). \tag{4.1}$$

Because the players make their selections independently, this simplifies to

$$J = \sum_{i,j} a_{ij} \text{ Prob}(\mathsf{P}_1 \text{ selects } i) \text{ Prob}(\mathsf{P}_2 \text{ selects } j) = \sum_{i,j} a_{ij} y_i z_j = y'Az,$$

where y and z are the following column vectors:

$$y := \begin{bmatrix} y_1 \\ y_2 \\ \vdots \\ y_m \end{bmatrix}, \qquad z := \begin{bmatrix} z_1 \\ z_2 \\ \vdots \\ z_n \end{bmatrix}.$$

Objective (mixed policies). The player P_1 wants to minimize the expected outcome $J = y'Az$ and the player P_2 and wants to maximize the same quantity.

There are two common interpretations for mixed policies:

- In the *repeated game paradigm*, the same two players face each other multiple times. In each game they choose their actions randomly according to preselected mixed policies (independently from each other and independently from game to game) and their goal is to minimize/maximize the cost/reward averaged over all the games played.

  This paradigm makes sense in many games in economics, e.g., in the advertising campaign game or the tax-payers auditing game; and also in political/social "engineering," e.g., in the crime/war deterrence games or the worker's compensation game.

- In the *large population paradigm*, there is large population of players P_1 and another equally large population of players P_2. All players only play pure policies, but the percentage of players that play each pure policy matches the probabilities of the mixed policies. Two players are then selected randomly from each population (independently) and they play against each other. The goal is to select a "good mix" for the populations so as to minimize/maximize the expected cost/reward.

  This paradigm also makes sense in some of the above examples, e.g., tax auditing, crime deterrence, or workers compensation. In addition, it makes sense to some robust design problems.

Notation. The *tax-payers auditing game* is played between the IRS and a tax-payer. The former has to decide if it is worth it to audit a tax-payer and the latter if it is worth it to cheat on the tax return.

Note. The *crime/war deterrence game* is just a high-stakes version of the tax-payers auditing game

4.2 Mixed Action Spaces

Consider a game specified by an $m \times n$ matrix A, with m actions for P_1 and n actions for P_2. When considering pure policies, the *(pure) action spaces* for P_1 and P_2 consist of the following finite sets

$$\{1, 2, \dots, m\} \quad \text{and} \quad \{1, 2, \dots, n\}$$

respectively. With *mixed policies*, the players choose distributions so the *(mixed) action spaces* for P_1 and P_2 consist of the following (infinite) continuous sets

$$\mathcal{Y} := \left\{ y \in \mathbb{R}^m : \sum_i y_i = 1,\ y_i \geq 0,\ \forall i \right\}, \qquad \mathcal{Z} := \left\{ z \in \mathbb{R}^n : \sum_j z_j = 1,\ z_j \geq 0,\ \forall j \right\},$$

respectively.

Attention! Pure policies still exist within the mixed action spaces. For example, the vector

$$[0 \quad 1 \quad 0 \quad \cdots \quad 0]' \in \mathcal{Y}$$

corresponds to the pure policy that consists of picking action 2, because we have that the probability y_2 of picking this action is one, whereas the probabilities y_i, $i \neq 2$ of picking other actions are all equal to zero.

Note. We recall that the set of actions available to a particular player is called their *action space*. With pure policies, each player has a finite number of choices and thus a finite action space.

Note. With mixed policies, each player has a continuum of possible actions and thus an action space with infinitely many elements.

Notation. Sets such as $\mathcal{Y}$ and $\mathcal{Z}$ are called *(probability) simplexes*.

4.3 Mixed Security Policies and Saddle-point Equilibrium

By introducing mixed policies, we essentially enlarged the action spaces for both players and we will see shortly that a consequence of this enlargement is that saddle-point equilibria now always exist. However, before that we need to extend the notions of security policies and saddle-point equilibria to the new type of policies, which is quite straightforward.

Definition 4.1 (Mixed security policy). Consider a matrix game defined by the matrix A. The *average security level* for P_1 (the minimizer) is defined by

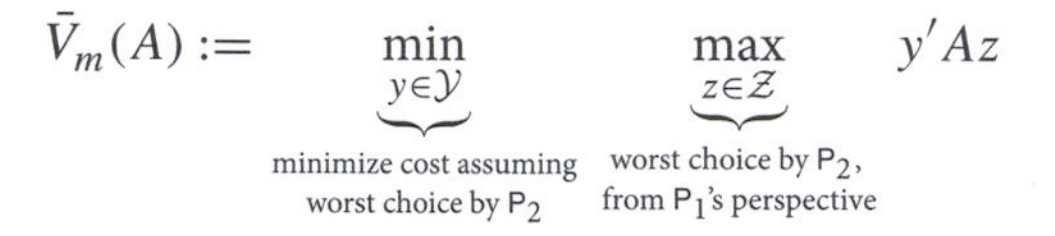

$$\bar{V}_m(A) := \underbrace{\min_{y\in\mathcal{Y}}}_{\substack{\text{minimize cost assuming}\\ \text{worst choice by } \mathsf{P}_2}} \underbrace{\max_{z\in\mathcal{Z}}}_{\substack{\text{worst choice by } \mathsf{P}_2,\\ \text{from } \mathsf{P}_1\text{'s perspective}}} y'Az$$

Notation. Equation (4.2) is often written as $y^* \in \arg\min_y \max_z y'Az$. The use of "$\in$" instead of "$=$" emphasizes that there may be several y^* that achieve the minimum.

and the corresponding *mixed security policy* for P_1 is any y^* that achieves the desired average security level, i.e.,

$$\underbrace{\max_{z\in\mathcal{Z}} y^{*\prime}Az = \bar{V}_m(A)}_{y^* \text{ achieves the minimum}} := \min_{y\in\mathcal{Y}} \max_{z\in\mathcal{Z}} y'Az. \tag{4.2}$$

Conversely, the *average security level* for P_2 (the maximizer) is defined by

$$\underline{V}_m(A) := \underbrace{\max_{z\in\mathcal{Z}}}_{\substack{\text{maximize reward assuming}\\ \text{worst choice by } \mathsf{P}_1}} \underbrace{\min_{y\in\mathcal{Y}}}_{\substack{\text{worst choice by } \mathsf{P}_1,\\ \text{from } \mathsf{P}_2\text{'s perspective}}} y'Az.$$

Notation. Equation (4.3) is often written as $z^* \in \arg\max_z \min_y y'Az$. The use of "$\in$" instead of "$=$" emphasizes that there may be several y^* that achieve the minimum.

and the corresponding *mixed security policy* for P_2 is any z^* that achieves the desired average security level, i.e.,

$$\underbrace{\min_{y\in\mathcal{Y}} y'Az^* = \underline{V}_m(A)}_{z^* \text{ achieves the maximum}} := \max_{z\in\mathcal{Z}} \min_{y\in\mathcal{Y}} y'Az. \tag{4.3}$$

Notation. A mixed saddle-point equilibria is often also called a *saddle-point equilibrium in mixed policies.*

Definition 4.2 (Mixed saddle-point equilibrium). Consider a matrix game defined by the matrix A. A pair of policies $(y^*, z^*) \in \mathcal{Y} \times \mathcal{Z}$ is called a *mixed saddle-point equilibrium* if

$$y^{*\prime}Az^* \le y'Az^*, \quad \forall y \in \mathcal{Y} \tag{4.4}$$

$$y^{*\prime}Az^* \ge y^{*\prime}Az, \quad \forall z \in \mathcal{Z} \tag{4.5}$$

Notation. The terminology "saddle point" arises from the fact that $y^{*\prime}Az^*$ increases along the "y-direction" and decreases along the "z-direction." With some imagination this looks like a horse's saddle.

and $y^{*\prime}Az^*$ is called the *mixed saddle-point value*. Equations (4.4)–(4.5) are often re-written as

$$y^{*\prime}Az \le y^{*\prime}Az^* \le y'Az^*, \quad \forall y \in \mathcal{Y}, z \in \mathcal{Z}.$$

Average security levels/policies satisfy the following three properties:

Proposition 4.1 (Security levels/policies). For every (finite) matrix A, the following properties hold:

P4.1 Average security levels are well defined and unique.

P4.2 Both players have mixed security policies (not necessarily unique).

P4.3 The average security levels always satisfy the following inequalities:

Note. An important consequence of this is that when there is a pure saddle-point equilibrium, $\underline{V}(A) = \bar{V}(A)$ are equal and therefore the average security levels are exactly the same as the (pure) security levels.

$$\underbrace{\underline{V}(A)}_{\max_j \min_i a_{ij}} \leq \underbrace{\underline{V}_m(A)}_{\max_z \min_y y'Az} \leq \underbrace{\bar{V}_m(A)}_{\min_y \max_z y'Az} \leq \underbrace{\bar{V}(A).}_{\min_i \max_j a_{ij}} \tag{4.6}$$

□

The inequality (4.6) expresses a perhaps unexpected feature of mixed policies: they lead to security levels that are "better" than those of pure policies *for both players*. In particular, the left-most inequality in (4.6) means that the mixed security level for the maximizer is larger than the pure security level and the right-most inequality in (4.6) means that the mixed security level for the minimizer is smaller than the pure security level.

Properties P4.1–P4.3 are only slightly less trivial to prove than the corresponding properties P3.1–P3.3 that we encountered before for pure security levels. Take for example

$$\underline{V}_m(A) = \max_{z \in \mathcal{Z}} \min_{y \in \mathcal{Y}} y'Az. \tag{4.7}$$

The property P4.1 implicitly states that the min and max in the definition of $\underline{V}_m(A)$ are achieved at specific points in $\mathcal{Y}$ and $\mathcal{Z}$, respectively. It turns out that this is true because we are minimizing/maximizing a continuous function over a compact set (i.e., bounded and closed) and Weierstrass' Theorem guarantees that such a minimum/maximum always exists at some point in the set.

Note 5. We are actually applying Weierstrass' Theorem twice. ▷ p. 40

Since the max in (4.7) is achieved at some point $z^* \in \mathcal{Z}$, we automatically have that z^* can be used in a security policy for P_2 (the maximizer), which justifies P4.2. The same exact reasoning applied for $\bar{V}_m(A)$ and the corresponding mixed security policy.

The inequalities in P4.3 are also fairly straightforward to prove. We start with the one in the left-hand side:

Note. Here we are basically using the fact that pure policies still exist within the mixed action spaces.

$$\underline{V}_m(A) := \max_{z \in \mathcal{Z}} \min_{y \in \mathcal{Y}} y'Az \geq \max_{z \in \{e_1, e_2, \dots, e_n\}} \min_{y \in \mathcal{Y}} y'Az = \max_j \min_{y \in \mathcal{Y}} y' \underbrace{Ae_j}_{j\text{th column of } A}$$

where $\{e_1, e_2, \dots, e_n\} \subset \mathcal{Z}$ are the canonical basis of $\mathbb{R}^n$. But then

$$\underline{V}_m(A) \geq \max_j \min_{y \in \mathcal{Y}} y' \begin{bmatrix} a_{1j} \\ a_{2j} \\ \vdots \\ a_{mj} \end{bmatrix} = \max_j \min_{y \in \mathcal{Y}} \sum_{i=1}^{m} y_i a_{ij}.$$

We now use an equality that is useful whenever we need to optimize a linear function over a simplex:

Note 6. This result will be used numerous time throughout the course. ▷ p. 40

$$\min_{y \in \mathcal{Y}} \sum_{i=1}^{m} y_i a_{ij} = \min_i a_{ij}.$$

This equality holds because the minimum is achieved by placing all the probability weight at the value of i for which the constant a_{ij} is the smallest. We therefore conclude that

$$\underline{V}_m(A) \geq \max_j \min_i a_{ij} =: \underline{V}(A),$$

which is the left-most inequality in P4.3.

The right-most inequality can be proved in an analogous fashion, but starting with $\bar{V}_m(A)$, instead of $\underline{V}_m(A)$. Finally, the middle inequality can be proved in exactly the same way that we used to prove that $\underline{V}(A) \leq \bar{V}(A)$:

$$\underline{V}_m(A) = \underbrace{\min_y y'Az^*}_{\substack{\text{where } z^* \text{ is a mixed} \\ \text{security policy}}} \leq \min_y \max_z y'Az =: \bar{V}_m(A).$$

Note 5. To prove property P4.1, we need to apply Weierstrass' Theorem twice.

1. First note that $y'Az$ is a continuous function of y (for each fixed z), which is being minimized over the compact set $\mathcal{Y}$. By Weierstrass' Theorem, there exists a point $y^* \in \mathcal{Y}$ at which the minimum is achieved, i.e.,

$$y^{*\prime}Az = \min_{y\in\mathcal{Y}} y'Az =: f(z).$$

Moreover the value of the minimum $f(z)$ is a continuous function of z.

2. Second we use the fact that $f(z)$ is continuous and is being maximized over the compact set $\mathcal{Z}$. Again by the Weierstrass' Theorem, there exists a point $z^* \in \mathcal{Z}$ at which the maximum is achieved, i.e.,

$$f(z^*) = \max_{z\in\mathcal{Z}} f(z) = \max_{z\in\mathcal{Z}} \min_{y\in\mathcal{Y}} y'A\,z.$$ □

Note 6 (Optimization of linear functions over simplexes). The following elementary result, which was used above, will be used numerous time throughout this book. We state it here in the form of a Lemma for ease of reference. □

Lemma 4.1 (Optimization of linear functions over simplexes). Consider a probability simplex

$$\mathcal{X} := \left\{ x \in \mathbb{R}^m : \sum_i x_i = 1,\, x_i \geq 0,\, \forall i \right\},$$

and a linear function f of the form

$$f(x) = \sum_{i=1}^{m} a_i x_i.$$

Then

$$\min_{x\in\mathcal{X}} f(x) = \min_{i\in\{1,2,\ldots,m\}} a_i, \qquad \max_{x\in\mathcal{X}} f(x) = \max_{i\in\{1,2,\ldots,m\}} a_i.$$ □

4.4 Mixed Saddle-Point Equilibrium vs. Average Security Levels

Just as with pure policies, the existence of a mixed saddle-point equilibrium is closely related to the average security levels for the two players. In fact, one can prove the following result in a similar way to the analogous result for pure policies (see Theorem 3.1):

Theorem 4.1 (Mixed saddle-point equilibrium vs. security levels). A matrix game defined by A has a mixed saddle-point equilibrium *if and only if*

Notation. In short, the min and max commute.

$$\underline{V}_m(A) := \max_{z \in \mathcal{Z}} \min_{y \in \mathcal{Y}} y'Az = \min_{y \in \mathcal{Y}} \max_{z \in \mathcal{Z}} y'Az =: \bar{V}_m(A). \tag{4.8}$$

In particular,

1. if (y^*, z^*) is a mixed saddle-point equilibrium then y^* and z^* are mixed security policies for P_1 and P_2, respectively and (4.8) is equal to the mixed saddle-point value.
2. if (4.8) holds and y^* and z^* are mixed security policies for P_1 and P_2, respectively then (y^*, z^*) is a mixed saddle-point equilibrium and its value is equal to (4.8). □

Notation. An important consequence of this is that all mixed saddle-point equilibria must have exactly the same mixed saddle-point values, which is called the *value of the game* and is denoted by $V_m(A)$.

However, there is a key difference between pure and mixed policies: it turns out that (4.8) holds *for every matrix A*. This fact, which is known as the Minimax Theorem, will be proved in Lecture 5.

Example 4.1 (Network routing game). Recall the network routing game introduced in Section 1.4, but for the simple network in Figure 4.1. This game can be viewed as a matrix game between the router P_1 and the attacker P_2 if we make the following associations:

$$\mathsf{P}_1\text{'s actions: } \begin{cases} \text{send packet through link 1} & \equiv 1 \\ \text{send packet through link 2} & \equiv 2 \end{cases} \qquad \mathsf{P}_2\text{'s actions: } \begin{cases} \text{attack link 1} & \equiv 1 \\ \text{attack link 2} & \equiv 2 \end{cases}$$

$$\text{outcomes: } \begin{cases} \text{packet arrives} & \equiv -1 \quad \mathsf{P}_1 \text{ wins} \\ \text{packet is intercepted} & \equiv +1 \quad \mathsf{P}_2 \text{ wins} \end{cases}$$

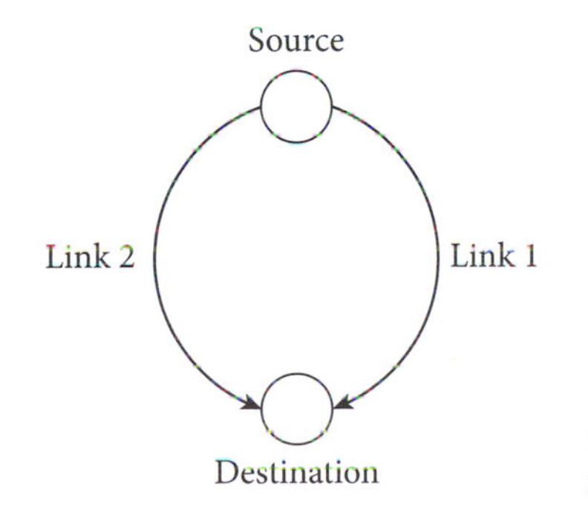

Figure 4.1 Network routing game in a two-link network.

which leads to the following matrix

$$A = \underbrace{\begin{bmatrix} +1 & -1 \\ -1 & +1 \end{bmatrix}}_{\mathsf{P}_2 \text{ choices}} \Big\} \mathsf{P}_1 \text{ choices}$$

with pure security levels

$$\underline{V}(A) = -1, \qquad \bar{V}(A) = +1,$$

showing that there are no saddle-point equilibria in pure policies. For mixed policies, we have

$$\begin{aligned} \underline{V}_m(A) &= \max_{z \in \mathcal{Z}} \min_{y \in \mathcal{Y}} y'Az = \max_{z \in \mathcal{Z}} \min_{y \in \mathcal{Y}} y_1(z_1 - z_2) + y_2(z_2 - z_1) \\ &= \max_{z \in \mathcal{Z}} \min\{z_1 - z_2, z_2 - z_1\}, \end{aligned}$$

where we used Lemma 4.1. To compute the maximum over z, we note that

$$\min\{z_1 - z_2, z_2 - z_1\} \begin{cases} = 0 & z_1 = z_2 \\ < 0 & z_1 \neq z_2. \end{cases}$$

So the maximum over z is obtained for $z_1 = z_2 = \frac{1}{2}$, which leads to

$$\underline{V}_m(A) = 0,$$

with a mixed security policy for P_2 given by $z^* := [\, \frac{1}{2} \;\; \frac{1}{2} \,]$. As for the other security levels, we have

$$\begin{aligned} \bar{V}_m(A) &= \min_{y \in \mathcal{Y}} \max_{z \in \mathcal{Z}} y'Az = \min_{y \in \mathcal{Y}} \max_{z \in \mathcal{Z}} z_1(y_1 - y_2) + z_2(y_2 - y_1) \\ &= \min_{y \in \mathcal{Y}} \max\{y_1 - y_2, y_2 - y_1\} = 0, \end{aligned}$$

with a mixed security policy for P_1 given by $y^* := [\, \frac{1}{2} \;\; \frac{1}{2} \,]$.

Not surprisingly, we conclude that $\underline{V}_m(A) = \bar{V}_m(A) = 0$ and therefore we conclude from Theorem 4.1 that this game has a mixed saddle-point equilibrium (y^*, z^*), which basically consists of each player selecting each of the links with equal probabilities. □

Example 4.2 (Rock-paper-scissors). Recall the rock-paper-scissors game introduced in Section 4.1, with matrix representation

$$A = \underbrace{\begin{bmatrix} 0 & +1 & -1 \\ -1 & 0 & +1 \\ +1 & -1 & 0 \end{bmatrix}}_{\mathsf{P}_2 \text{ choices}} \Bigg\} \mathsf{P}_1 \text{ choices}.$$

We had seen before that for this game $\underline{V}(A) = -1$ and $\bar{V}(A) = 1$. For mixed policies, using Lemma 4.1, we conclude that

$$\underline{V}_m(A) = \max_{z \in \mathcal{Z}} \min_{y \in \mathcal{Y}} y'Az = \max_{z \in \mathcal{Z}} \min_{y \in \mathcal{Y}} y_1(z_2 - z_3) + y_2(z_3 - z_1) + y_3(z_1 - z_2)$$
$$= \max_{z \in \mathcal{Z}} \min\{z_2 - z_3, z_3 - z_1, z_1 - z_2\},$$

Note. Why? To get the maximum over z we must have $z_2 \geq z_3$ since otherwise $z_2 - z_3 < 0$ and we would get the minimum smaller than zero. For similar reasons, we must also have $z_2 \geq z_1$ and $z_1 \geq z_2$. The only way to satisfy these three inequalities simultaneously is to have all z_j equal to each other.

which is maximized with $z_1 = z_2 = z_3 = \frac{1}{3}$. This means that

$$V_m(A) = 0,$$

corresponding to a mixed security policy for P_2 given by $z^* := [\, \frac{1}{3} \;\; \frac{1}{3} \;\; \frac{1}{3} \,]$. Similarly,

$$\bar{V}(A) = \min y \in \mathcal{Y} \max_z y'Az = \cdots = 0,$$

with a mixed security policy for P_1 given by $y^* := [\, \frac{1}{3} \;\; \frac{1}{3} \;\; \frac{1}{3} \,]$.

Note. A lot of work to conclude what every 7 year old learns in the school yard. Starting in Lecture 6, we will discuss more systematic ways to compute mixed security policies, but some of the key ideas are already captured by these examples.

Not surprisingly, we conclude that $\underline{V}_m(A) = \bar{V}_m(A) = 0$ and therefore this game has a mixed saddle-point equilibrium (y^*, z^*) that basically consists of each player selecting rock, paper, or scissors with equal probabilities. □

4.5 General Zero-Sum Games

In subsequent lectures we will continue to enlarge the types of games considered and the action spaces for the players. It is therefore convenient to formalize several of the concepts and results that we have encountered so far to more general classes of games.

Consider a general two-player zero-sum game G in which the players P_1 and P_2 are allowed to select policies within *action spaces* Γ_1 and Γ_2, respectively. In general, action spaces can be finite as in matrix games with pure policies or infinite as in matrix games with mixed policies. However, we do not restrict the action spaces to these two examples and we will shortly find other types of action spaces.

Attention! We now allow the outcome J to depend on the policies in an arbitrary fashion.

For a particular pair of policies $\gamma \in \Gamma_1, \sigma \in \Gamma_2$ we denote by $J(\gamma, \sigma)$ the *outcome of the game* when player P_1 uses policy γ and P_2 uses policy σ. As usual, P_1 wants to minimize the outcome $J(\gamma, \sigma)$, whereas P_2 wants to maximize it.

With so few assumptions on the action spaces and outcomes, we need to slightly adapt our previous definitions of security policies and levels:

Notation 1 The *infimum* of a set is its largest lower bound and the *supremum* of a set is its smallest upper bound.. ▷ p. 44

Definition 4.3 (Security policy). The *security level* for P_1 (the minimizer) is defined by

$$\bar{V}_{\Gamma_1,\Gamma_2}(G) := \underbrace{\inf_{\gamma \in \Gamma_1}}_{\substack{\text{minimize cost assuming} \\ \text{worst choice by } P_2}} \underbrace{\sup_{\sigma \in \Gamma_2}}_{\substack{\text{worst choice by } P_2, \\ \text{from } P_1\text{'s perspective}}} J(\gamma, \sigma)$$

Notation. Equation (4.9) is often written as $\gamma^* \in \arg\min_\gamma \sup_\sigma J(\gamma, \sigma)$. The use of "$\in$" instead of "$=$" emphasizes that there may be several (or none) γ^* that achieve the infimum.

and a *security policy* for P_1 is any policy $\gamma^* \in \Gamma_1$ for which the infimum above is achieved, i.e.,

$$\underbrace{\sup_{\sigma \in \Gamma_2} J(\gamma^*, \sigma) = \bar{V}_{\Gamma_1,\Gamma_2}(G)}_{\gamma^* \text{ achieves the infimum}} := \inf_{\gamma \in \Gamma_1} \sup_{\sigma \in \Gamma_2} J(\gamma, \sigma). \tag{4.9}$$

Conversely, the *security level* for P_2 (the maximizer) is defined by

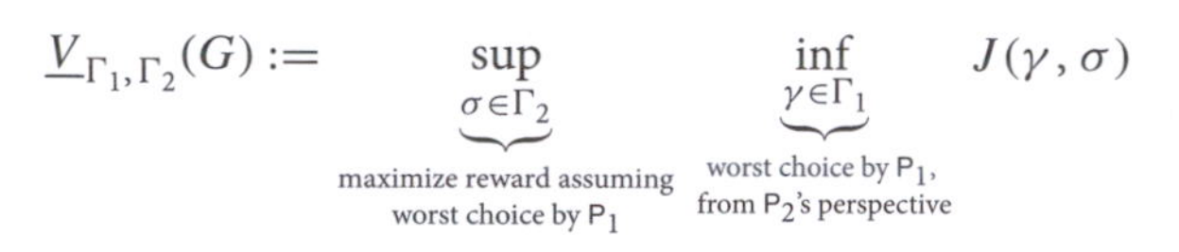

$$\underline{V}_{\Gamma_1,\Gamma_2}(G) := \underbrace{\sup_{\sigma\in\Gamma_2}}_{\text{maximize reward assuming worst choice by } \mathsf{P}_1} \underbrace{\inf_{\gamma\in\Gamma_1}}_{\text{worst choice by } \mathsf{P}_1, \text{ from } \mathsf{P}_2\text{'s perspective}} J(\gamma,\sigma)$$

and a *security policy* for P_2 is any policy $\sigma^* \in \Gamma_2$ for which the supremum above is achieved, i.e.,

Notation. Equation (4.10) is often written as $\sigma^* \in \arg\max_\sigma \inf_\gamma J(\gamma,\sigma)$. The use of "$\in$" instead of "$=$" emphasizes that there may be several (or none) σ^* that achieve the supremum.

$$\underbrace{\sup_{\gamma\in\Gamma_1} J(\gamma,\sigma^*) = \underline{V}_{\Gamma_1,\Gamma_2}(G)}_{\sigma^* \text{ achieves the maximum}} := \sup_{\sigma\in\Gamma_2}\inf_{\gamma\in\Gamma_1} J(\gamma,\sigma). \tag{4.10}$$

Note. These conditions are consistent with the Meta Definition 1.1.

Definition 4.4 (Saddle-point equilibrium). A pair of policies $(\gamma^*,\sigma^*) \in \Gamma_1 \times \Gamma_2$ is called a *saddle-point equilibrium* if

$$J(\gamma^*,\sigma^*) \le J(\gamma,\sigma^*), \quad \forall\gamma \in \Gamma_1 \tag{4.11a}$$

$$J(\gamma^*,\sigma^*) \ge J(\gamma^*,\sigma), \quad \forall\sigma \in \Gamma_2 \tag{4.11b}$$

and $J(\gamma^*,\sigma^*)$ is called the *saddle-point value*. Equations (4.11a)–(4.11b) are often re-written as

$$J(\gamma^*,\sigma) \le J(\gamma^*,\sigma^*) \le J(\gamma,\sigma^*), \quad \forall\gamma \in \Gamma_1, \sigma \in \Gamma_2.$$

Notation 1 (Infimum and supremum). The *infimum* of a set $\mathcal{S} \subset \mathbb{R}$ is the largest number y^* such that

$$y^* \le s, \quad \forall s \in \mathcal{S}. \tag{4.12}$$

When there is no number y^* that satisfies (4.12), the infimum is defined to be $-\infty$. Under this definition every set $\mathcal{S} \subset \mathbb{R}$ has an infimum, but the infimum may not belong to the set. When the infimum does belong to the set it is said to be a *minimum*. For example,

Notation. It is common to write (4.13) as $\inf_{x\ge 0} e^{-x} = 0$, $\inf_{x\le 0} e^{-x} = \min_{x\le 0} e^{-x} = 1$.

$$\inf\{e^{-x} : x \ge 0\} = 0, \qquad \inf\{e^{-x} : x \le 0\} = \min\{e^{-x} : x \le 0\} = 1. \tag{4.13}$$

Note that the infimum of the set in the left hand side is not a minimum since 0 does not belong to the set, but the infimum of the set in the right hand side does belong to the set.

The *supremum* of a set $\mathcal{S} \subset \mathbb{R}$ is the smallest number z^* such that

$$z^* \ge s, \quad \forall s \in \mathcal{S}. \tag{4.14}$$

When there is no number z^* that satisfies (4.14), the supremum is defined to be $+\infty$. Under this definition every set $\mathcal{S} \subset \mathbb{R}$ has a supremum. When the supremum belongs to the set it is said to be a *maximum*. □

Even with such little structure on the action spaces, security levels, policies, and saddle-point equilibria still satisfy very useful properties:

Proposition 4.2 (Security levels/policies/saddle-point equilibria). The following properties hold for every zero-sum game:

Note. However, security levels may take infinite values (including $+\infty$ or $-\infty$) and security policies may not exist.

P4.4 Security levels are well defined and unique.

P4.5 The security levels always satisfy the following inequality:

$$\underline{V}_{\Gamma_1,\Gamma_2}(G) := \sup_{\sigma\in\Gamma_2} \inf_{\gamma\in\Gamma_1} J(\gamma,\sigma) \le \inf_{\gamma\in\Gamma_1} \sup_{\sigma\in\Gamma_2} J(\gamma,\sigma) =: \bar{V}_{\Gamma_1,\Gamma_2}(G).$$

P4.6 When $\underline{V}_{\Gamma_1,\Gamma_2}(G) = \bar{V}_{\Gamma_1,\Gamma_2}(G)$ and there exist security policies γ^* and σ^* for P_1 and P_2, respectively then (γ^*, σ^*) is a saddle-point equilibrium and its value $J(\gamma^*, \sigma^*)$ is equal to $\underline{V}_{\Gamma_1,\Gamma_2}(G) = \bar{V}_{\Gamma_1,\Gamma_2}(G)$.

Notation. An important consequence of this is that all saddle-point equilibria must have exactly the same saddle-point values, which is called the *value of the game* and is denoted by $V_{\Gamma_1,\Gamma_2}(G)$.

P4.7 If there exists a saddle-point equilibrium (γ^*, σ^*) then γ^* and σ^* are security policies for P_1 and P_2, respectively and $\underline{V}_{\Gamma_1,\Gamma_2}(G) = \bar{V}_{\Gamma_1,\Gamma_2}(G) = J(\gamma^*, \sigma^*)$.

P4.8 If (γ_1^*, σ_1^*) and (γ_2^*, σ_2^*) are both saddle-point equilibria then (γ_1^*, σ_2^*) and (γ_2^*, σ_1^*) are also saddle-point equilibria and all these equilibria have exactly the same value. □

Example 4.3 (Resistive circuit design). Recall the robust design problem for a resistive circuit discussed in Section 1.3.

- P_1 is the circuit designer and picks the nominal resistance R_{nom} to *minimize* the current error

$$e(R_{\text{nom}}, \delta) = \left|\frac{1}{R} - 1\right| = \left|\frac{1}{(1+\delta)R_{\text{nom}}} - 1\right|.$$

Note. Robust designs generally lead to zero-sum games, such as this one.

- P_2 is "nature" and picks the value of $\delta \in [-0.1, 0.1]$ to *maximize* the same current error e.

We have seen before that the policy

$$\pi_1^* : \mathsf{P}_1 \text{ selects } R_{\text{nom}} = \frac{100}{99} \tag{4.15}$$

is a security policy for P_1 and leads to a security level $\bar{V} = 0.1$. Since for any given value of $\delta \in [-0.1, +0.1]$, P_1 could get the error to be exactly equal to 0 by an appropriate choice of R_{nom}. The security level for P_2 is $\underline{V} = 0$ and any policy is a security policy for P_2. In view of the fact that

$$\underline{V} = 0 < \bar{V} = 0.1,$$

Note. This is not a matrix game like the ones we have seen so far since each player has infinitely many choices and we are selecting a distribution for P_2 over an interval, but it still fits within the general framework we have been discussing.

we conclude that this game does not have a saddle-point equilibrium. Suppose now that resistors are being drawn from a box that contains

$$\begin{cases} 45\% \text{ of the resistors with } \delta = -0.1 \\ 55\% \text{ of the resistors with } \delta = +0.1 \end{cases}$$

In this case, when P_1 uses the policy π_1 and a resistor is drawn randomly from the box, the expected value of the error is

$$E[e] = 0.45 \times \left| \frac{1}{(1-0.1)\frac{100}{99}} - 1 \right| + 0.55 \times \left| \frac{1}{(1+0.1)\frac{100}{99}} - 1 \right| = 0.1.$$

We can view this box of resistors as P_2's mixed policy, by imagining that nature picked the distribution of resistors in the box in order to maximize the expected value of the error. This corresponds to the following mixed policy for P_2

$$\pi_2^* : \mathsf{P}_2 \text{ selects } \begin{cases} -0.1 & \text{with probability } 0.45 \\ +0.1 & \text{with probability } 0.55. \end{cases} \tag{4.16}$$

We can state the following:

1. Restricting our attention to *pure policies*, the security levels of the game are

$$\underline{V} = 0 < \bar{V} = 0.1$$

The policy π_1^* in (4.15) is a security policy for P_1 and any policy for P_2 is a security policy for this player. The gap between the security levels indicates that there are no pure saddle points.

2. Enlarging the universe of policies to consider *mixed policies*, the average security levels of the game become

$$\underline{V}_m = \bar{V}_m = 0.1$$

The policy π_1^* in (4.15) is still a mixed security policy for P_1, but now π_2^* in (4.16) is a mixed security policy for P_2. The equality between the security levels indicates that (π_1^*, π_2^*) is a mixed saddle-point equilibrium.

Note. In essence, by introducing mixed policies, we allowed P_2 to raise her security level to 0.1.

Note. See Exercise 4.1. ▷ p. 47

Note. To regard π_1^* in (4.15) as a mixed policy, we should think of this policy as the probability distribution for R_{nom} that places all probability mass at $\frac{100}{99}$.

Note. It is very convenient that the security policy for the designer is a pure policy because otherwise there was not clear "best" decision.

Note. This situation should be contrasted with the Rock-Papers-Scissors game, where there was also a gap between the pure security levels, but this gap could not be closed while keeping any of the pure security policies. In the Rock-Papers-Scissors game, the pure security policies are essentially useless as they can only "guarantee" that a player will do no worse than losing (see Section 4.1). ▷ p. 35

The fact that $R_{\text{nom}} = \frac{100}{99}$ is part of a mixed saddle-point equilibrium, tells us that this design is not conservative in the sense that:

1. our selection of $R_{\text{nom}} = \frac{100}{99}$ guarantees that the current error remains below 0.1 for any resistor R with error below 10%, and
2. there actually exists a population of resistors (with error below 10%) so that by extracting R randomly from this population, we do get $\mathrm{E}[e] = 0.1$.

If the security policy $R_{\text{nom}} = \frac{100}{99}$ was *not* a mixed saddle-point equilibrium then there would be no underlying distribution of resistors that could lead to $\mathrm{E}[e] = 0.1$. This would indicate that by choosing $R_{\text{nom}} = \frac{100}{99}$ we were "protecting" our design against a "phantom" worst-case distribution of resistances that actually did not exist. □

4.6 Practice Exercises

4.1 (Resistive circuit design). Consider the robust design problem for a resistive circuit discussed in Section 1.3 and Examples 4.3. Show that the average security levels of this game are given by

$$\underline{V}_m = \bar{V}_m = 0.1$$

and that the policies (π_1^*, π_2^*) defined by

$$\pi_1^*: \mathsf{P}_1 \text{ selects } R_{\text{nom}} = \frac{100}{99}$$

$$\pi_2^*: \mathsf{P}_2 \text{ selects } \begin{cases} -0.1 & \text{with probability } 0.45 \\ +0.1 & \text{with probability } 0.55 \end{cases}$$

form a mixed saddle-point equilibrium.

This problem is a little more involved than you might think. Hint: Start by showing that π_1^* is a pure security policy for P_1, leading to $\bar{V} = 0.1$. Then verify that π_1^* and π_2^* satisfy the general conditions (4.11a)–(4.11b) for a mixed saddle-point equilibrium.

Solution to Exercise 4.1. To show that $R_{\text{nom}} = \frac{100}{99}$ is a pure security policy for P_1 with security level equal to 0.1, we start by computing

$$\begin{aligned}
\bar{V} &= \min_{R_{\text{nom}} \ge 0} \max_{\delta \in [-0.1, 0.1]} \left| \frac{1}{(1+\delta) R_{\text{nom}}} - 1 \right| \\
&= \min_{R_{\text{nom}} \ge 0} \max_{\delta \in [-0.1, 0.1]} \begin{cases} \frac{1}{(1+\delta) R_{\text{nom}}} - 1 & \delta \le \frac{1}{R_{\text{nom}}} - 1 \\ 1 - \frac{1}{(1+\delta) R_{\text{nom}}} & \delta > \frac{1}{R_{\text{nom}}} - 1. \end{cases}
\end{aligned}$$

Since the top branch is monotone decreasing with respect to δ and the bottom branch is monotone increasing, the inner maximization is achieved at the extreme points for δ and we conclude that

$$\begin{aligned}
\bar{V} &= \min_{R_{\text{nom}} \ge 0} \begin{cases} \max\left\{ \frac{1}{(1-0.1) R_{\text{nom}}} - 1, 1 - \frac{1}{(1+0.1) R_{\text{nom}}} \right\} & -0.1 \le \frac{1}{R_{\text{nom}}} - 1, \, 0.1 > \frac{1}{R_{\text{nom}}} - 1 \\ \frac{1}{(1-0.1) R_{\text{nom}}} - 1 & -0.1 \le \frac{1}{R_{\text{nom}}} - 1, \, 0.1 \le \frac{1}{R_{\text{nom}}} - 1 \\ 1 - \frac{1}{(1+0.1) R_{\text{nom}}} & -0.1 > \frac{1}{R_{\text{nom}}} - 1, \, 0.1 > \frac{1}{R_{\text{nom}}} - 1 \end{cases} \\
&= \min_{R_{\text{nom}} \ge 0} \begin{cases} \max\left\{ \frac{1}{(1-0.1) R_{\text{nom}}} - 1, 1 - \frac{1}{(1+0.1) R_{\text{nom}}} \right\} & R_{\text{nom}} \in \left(\frac{10}{11}, \frac{10}{9} \right] \\ \frac{1}{(1-0.1) R_{\text{nom}}} - 1 & R_{\text{nom}} \le \frac{10}{11} \\ 1 - \frac{1}{(1+0.1) R_{\text{nom}}} & R_{\text{nom}} > \frac{10}{9}. \end{cases}
\end{aligned}$$

Considering the three branches separately, we conclude that the optimum is achieved at the top branch when

$$\frac{1}{(1-0.1)R_{\text{nom}}}-1=1-\frac{1}{(1+0.1)R_{\text{nom}}} \quad\Leftrightarrow\quad R_{\text{nom}}=\frac{100}{99}.$$

This shows that π_1^* is a security policy for P_1, corresponding to a security level of

$$\bar{V}=0.1.$$

To prove that (π_1, π_2) is a mixed saddle-point equilibrium we show that π_1^* and π_2^* satisfy the general conditions (4.11a)–(4.11b) for a mixed saddle-point equilibrium. We start by noting that since π_1^* is a pure security policy for P_1 with security level equal to 0.1, when P_1 uses this policy the error in the current satisfies

$$e\left(\frac{100}{99},\delta\right)=\left|\frac{1}{(1+\delta)\frac{100}{99}}-1\right|\leq 0.1, \quad \forall\delta\in[-0.1,0.1].$$

This means that, for any mixed policy π_2 for P_2, we must necessarily have

$$\mathrm{E}_{\pi_1^*\pi_2}\left[e\left(\frac{100}{99},\delta\right)\right]\leq 0.1, \quad \forall\pi_2, \tag{4.17}$$

where the subscript $_{\pi_1^*\pi_2}$ in the expected value means that we are fixing the (pure) policy π_1^* and taking δ to be a random variable with distribution π_2.

Suppose now that we fix P_2's policy to be π_2^* and consider an arbitrary policy π_1 for P_1, corresponding to a distribution for R_{nom} with probability density function $f(r)$. Then

$$\begin{aligned}
\mathrm{E}_{\pi_1,\pi_2^*} &= 0.45\int_0^\infty e(r,-0.1)f(r)dr+0.55\int_0^\infty e(r,0.1)f(r)dr \\
&= \int_0^{\frac{10}{9}} 0.45\left(\frac{1}{(1-0.1)r}-1\right)f(r)dr+\int_{\frac{10}{9}}^\infty 0.45\left(1-\frac{1}{(1-.01)r}\right)f(r)dr \\
&\quad+\int_0^{\frac{10}{11}} 0.55\left(\frac{1}{(1+0.1)r}-1\right)f(r)dr+\int_{\frac{10}{11}}^\infty 0.55\left(1-\frac{1}{(1+0.1)r}\right)f(r)dr \\
&= \int_0^{\frac{10}{11}}\left(0.45\left(\frac{1}{(1-0.1)r}-1\right)+0.55\left(\frac{1}{(1+0.1)r}-1\right)\right)f(r)dr \\
&\quad+\int_{\frac{10}{11}}^{\frac{10}{9}}\left(0.45\left(\frac{1}{(1-0.1)r}-1\right)+0.55\left(1-\frac{1}{(1+0.1)r}\right)\right)f(r)dr \\
&\quad+\int_{\frac{10}{9}}^\infty\left(0.45\left(1-\frac{1}{(1-0.1)r}\right)+0.55\left(1-\frac{1}{(1+0.1)r}\right)\right)f(r)dr \\
&= \int_0^{\frac{10}{11}}\left(\frac{1}{r}-1\right)f(r)dr+\int_{\frac{10}{11}}^{\frac{10}{9}}\frac{1}{10}f(r)dr+\int_{\frac{10}{9}}^\infty\left(1-\frac{1}{r}\right)f(r)dr.
\end{aligned}$$

Since

$$r\leq\frac{10}{11}\Rightarrow\frac{1}{r}-1\geq\frac{11}{10}-1=\frac{1}{10}$$

and

$$r \geq \frac{10}{9} \Rightarrow 1 - \frac{1}{r} \geq 1 - \frac{9}{10} = \frac{1}{10},$$

we conclude that $\mathrm{E}_{\pi_1,\pi_2^*}$ is minimized by selecting all the probability mass for the distribution π_1 to be in the interval $\left[\frac{10}{11}, \frac{10}{9}\right]$, leading to

$$\max_{\pi_1} \mathrm{E}_{\pi_1,\pi_2^*} = \frac{1}{10} = 0.1 = \mathrm{E}_{\pi_1^*,\pi_2^*}.$$

This and (4.17) establish that π_1^* and π_2^* satisfy the general conditions (4.11a)–(4.11b) for a saddle-point equilibrium.

4.2 (Security levels/policies/saddle-point equilibria). Prove Proposition 4.2.

Hint: Get inspiration from the proofs of Propositions 3.1, 3.2, and Theorem 3.1. However, note that you need to prove that P4.5 holds even when security policies do not exist.

Solution to Exercise 4.2. Property P4.4 holds simply because infima and suprema always exist in $\mathbb{R} \cup \{-\infty, +\infty\}$.

As for property P4.5, pick a small positive number $\epsilon > 0$ and let σ^ϵ be a policy for the maximizer P_2 that "almost" gets to the supremum, i.e.,

$$\inf_{\gamma \in \Gamma_1} J(\gamma, \sigma^\epsilon) + \epsilon \geq \underline{V}_{\Gamma_1,\Gamma_2}(G) := \sup_{\sigma \in \Gamma_2} \inf_{\gamma \in \Gamma_1} J(\gamma, \sigma).$$

Such a policy must exist because if one could not find policies for the maximizer arbitrarily close to $\underline{V}_{\Gamma_1,\Gamma_2}(G)$, then this would not be the supremum. Since

$$J(\gamma, \sigma^\epsilon) \leq \sup_{\sigma \in \Gamma_2} J(\gamma, \sigma)$$

we conclude that

$$\underline{V}_{\Gamma_1,\Gamma_2}(G) - \epsilon \leq \inf_{\gamma \in \Gamma_1} J(\gamma, \sigma^\epsilon) \leq \inf_{\gamma \in \Gamma_1} \sup_{\sigma \in \Gamma_2} J(\gamma, \sigma) =: \bar{V}_{\Gamma_1,\Gamma_2}(G).$$

Since this inequality must hold regardless of how small ϵ is, we conclude that P4.5 must hold.

To prove P4.6, we note that if γ^* is a security policy for P_1, we have that

$$\sup_{\sigma} J(\gamma^*, \sigma) = \bar{V}_{\Gamma_1,\Gamma_2}(G) \quad \Big(:= \inf_{\sigma} \sup_{\gamma} J(\gamma, \sigma)\Big)$$

and if σ^* is a security policy for P_2, we also have that

$$\inf_{\gamma} J(\gamma, \sigma^*) = \underline{V}_{\Gamma_1,\Gamma_2}(G) \quad \Big(:= \sup_{\sigma} \inf_{\gamma} J(\gamma, \sigma)\Big).$$

On the other hand, just because of what it means to be an inf/sup, we have that

$$\underline{V}_{\Gamma_1,\Gamma_2}(G) := \inf_{\gamma} J(\gamma, \sigma^*) \leq J(\gamma^*, \sigma^*) \leq \sup_{\sigma} J(\gamma^*, \sigma) =: \bar{V}_{\Gamma_1,\Gamma_2}(G).$$

When $\underline{V}_{\Gamma_1,\Gamma_2}(G) = \bar{V}_{\Gamma_1,\Gamma_2}(G)$ the inequalities above must actually be equalities and we obtain

$$J(\gamma^*, \sigma^*) = \sup_{\sigma} J(\gamma^*, \sigma) \Rightarrow J(\gamma^*, \sigma^*) \geq J(\gamma^*, \sigma), \quad \forall \sigma$$

$$J(\gamma^*, \sigma^*) = \inf_{\gamma} J(\gamma, \sigma^*) \Rightarrow J(\gamma^*, \sigma^*) \leq J(\gamma, \sigma^*), \quad \forall \gamma,$$

from which we conclude that (γ^*, σ^*) is indeed a saddle-point equilibrium.

To prove P4.7, we assume that (γ^*, σ^*) is a saddle-point equilibrium. In this case

$$J(\gamma^*, \sigma^*) \underbrace{=}_{\text{by (4.11a)}} \inf_{\gamma} J(\gamma, \sigma^*) \underbrace{\leq}_{\substack{\text{since } \sigma^* \text{ is one} \\ \text{particular } \sigma}} \sup_{\sigma} \inf_{\gamma} J(\gamma, \sigma) =: \underline{V}_{\Gamma_1,\Gamma_2}(G). \tag{4.18}$$

Similarly

$$J(\gamma^*, \sigma^*) \underbrace{=}_{\text{by (4.11b)}} \sup_{\sigma} J(\gamma^*, \sigma) \underbrace{\geq}_{\substack{\text{since } \gamma^* \text{ is one} \\ \text{particular } \gamma}} \inf_{\gamma} \sup_{\sigma} J(\gamma, \sigma) =: \bar{V}_{\Gamma_1,\Gamma_2}(G). \tag{4.19}$$

Therefore

$$\bar{V}_{\Gamma_1,\Gamma_2}(G) \leq J(\gamma^*, \sigma^*) \leq \underline{V}_{\Gamma_1,\Gamma_2}(G).$$

But since we already saw in P4.5 that $\bar{V}_{\Gamma_1,\Gamma_2}(G) \geq \underline{V}_{\Gamma_1,\Gamma_2}(G)$ all these inequalities are actually equalities, which proves P4.7. In addition, since (4.18) holds with equality, we conclude that σ^* must be a security policy for P_2 and since (4.19) holds with equality, γ^* must be a security policy for P_1.

Finally, to prove P4.8 we note that if (γ_1^*, σ_1^*) and (γ_2^*, σ_2^*) are both saddle-point equilibria then all four policies must be security policies and $\bar{V}_{\Gamma_1,\Gamma_2}(G) = \underline{V}_{\Gamma_1,\Gamma_2}(G) = J(\gamma_1^*, \sigma_1^*) = J(\gamma_2^*, \sigma_2^*)$, in view of P4.7. But then (γ_1^*, σ_2^*) and (γ_2^*, σ_1^*) are also saddle-point equilibria because of P4.6 with the same value.

4.7 Additional Exercise

4.3. In a *quadratic zero-sum game* the player P_1 selects a vector $u \in \mathbb{R}^{n_u}$, the player P_2 selects a vector $d \in \mathbb{R}^{n_d}$, and the corresponding outcome of the game is the quadratic form

$$J(u, d) := \begin{bmatrix} u' & d' & x' \end{bmatrix} \begin{bmatrix} P_{uu} & P_{ud} & P_{ux} \\ P'_{ud} & P_{dd} & P_{dx} \\ P'_{ux} & P'_{dx} & P_{xx} \end{bmatrix} \begin{bmatrix} u \\ d \\ x \end{bmatrix} \tag{4.20}$$

that P_1 wants to minimize and P_2 wants to maximize, where $x \in \mathbb{R}^n$ is a vector, $P_{uu} \in \mathbb{R}^{n_u \times n_u}$ a symmetric positive definite matrix, $P_{dd} \in \mathbb{R}^{n_d \times n_d}$ a symmetric negative definite matrix, and $P_{xx} \in \mathbb{R}^{n \times n}$ a symmetric matrix.

1. Show that

$$\begin{bmatrix} u^* \\ d^* \end{bmatrix} = - \begin{bmatrix} P_{uu} & P_{ud} \\ P'_{ud} & P_{dd} \end{bmatrix}^{-1} \begin{bmatrix} P_{ux} \\ P_{dx} \end{bmatrix} x \tag{4.21}$$

is the unique solution to the system of equations

$$\frac{\partial J(u^*, d^*)}{\partial u} = 0, \qquad \frac{\partial J(u^*, d^*)}{\partial d} = 0.$$

Hint: If the matrices A and $D - CA^{-1}B$ are invertible, then $\begin{bmatrix} A & B \\ C & D \end{bmatrix}$ *is also invertible.*

2. Show that the pair of policies $(u^*, d^*) \in \mathbb{R}^{n_u} \times \mathbb{R}^{n_d}$ is a saddle-point equilibrium for the quadratic zero-sum game with value

$$J(u^*, d^*) = x' \left(P_{xx} - \begin{bmatrix} P'_{ux} & P'_{dx} \end{bmatrix} \begin{bmatrix} P_{uu} & P_{ud} \\ P'_{ud} & P_{dd} \end{bmatrix}^{-1} \begin{bmatrix} P_{ux} \\ P_{dx} \end{bmatrix} \right) x.$$

LECTURE 5

Minimax Theorem

This lecture is devoted to the Minimax Theorem. This result was originally proved by von Neumann [15], who is generally recognized as the founder of Game Theory as a mathematical discipline. The proof included here is based on elementary results from convex analysis and appeared in a pioneering book by von Neumann and Morgenstern [11]. An historical perspective on the Minimax Theorem and the multiple techniques that can be used to prove it can be found in [5].

5.1 Theorem Statement

Consider a game specified by an $m \times n$ matrix A, with m actions for P_1 and n actions for P_2:

$$A = \underbrace{\begin{bmatrix} & \vdots & \\ \cdots & a_{ij} & \cdots \\ & \vdots & \end{bmatrix}}_{\mathsf{P}_2 \text{ choices (maximizer)}} \Big\} \mathsf{P}_1 \text{ choices (minimizer)}.$$

This lecture address the following fundamental theorem in matrix games:

Notation. In short, in mixed policies the min and max always commute.

Theorem 5.1 (Minimax). For every matrix A, the average security levels of both players coincide, i.e.,

$$\underline{V}_m(A) := \max_{z \in \mathcal{Z}} \min_{y \in \mathcal{Y}} y' A z = \min_{y \in \mathcal{Y}} \max_{z \in \mathcal{Z}} y' A z =: \bar{V}_m(A).$$

□

This result is sufficiently important that we will spend most of the lecture working on it.

We saw in Lecture 4 (Property P4.3) that

$$\underline{V}_m(A) \leq \bar{V}_m(A).$$

If the inequality were strict, there would be a constant c such that

$$\underline{V}_m(A) < c < \bar{V}_m(A).$$

The proof of this theorem consists in showing that this is not possible. We will show that for any number c, we either have

Note. This means that for every constant c, at least one of the players can guarantee a security level of c.

$$c \leq \underline{V}_m(A) \quad \text{or} \quad \bar{V}_m(A) \leq c.$$

However, to achieve this we need some terminology and a key result in convex analysis.

5.2 Convex Hull

Given k vectors $v_1, v_2, \ldots, v_k \in \mathbb{R}^n$, the *linear subspace generated by these vectors* is the set

$$\text{span}(v_1, v_2, \ldots, v_k) = \left\{ \sum_{i=1}^{k} \alpha_i v_i : \alpha_i \in \mathbb{R} \right\} \subset \mathbb{R}^n,$$

which is represented graphically in Figure 5.1. On the other hand, the *(closed) convex hull generated by these vectors* is the set

Notation. A linear combination with positive coefficients that add up to one, is called a *convex combination.*

$$\text{co}(v_1, v_2, \ldots, v_k) = \left\{ \sum_{i=1}^{k} \alpha_i v_i : \alpha_i \geq 0, \sum_{i=1}^{k} \alpha_i = 1 \right\} \subset \mathbb{R}^n,$$

which is represented graphically in Figure 5.2. The convex hull is always a *convex set* in the sense that

$$x_1, x_2 \in \text{co}(v_1, v_2, \ldots, v_k) \Rightarrow \frac{\lambda x_1 + (1-\lambda)x_2}{2} \in \text{co}(v_1, v_2, \ldots, v_k), \quad \forall \lambda \in [0, 1],$$

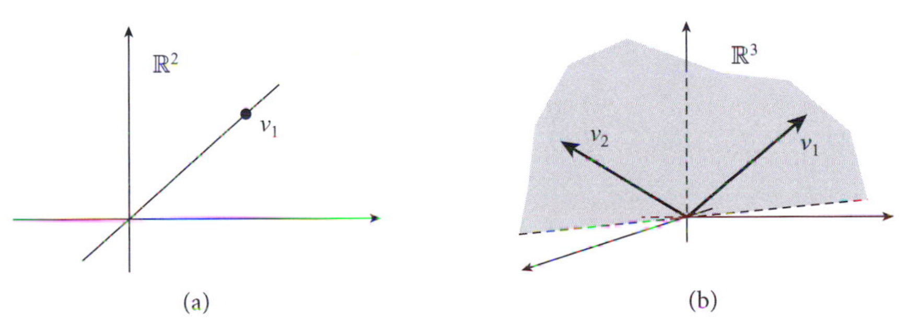

Figure 5.1 Linear subspaces. (a) Linear subspace of $\mathbb{R}^2$ generated by v_1; (b) Linear subspace of $\mathbb{R}^3$ generated by v_1, v_2.

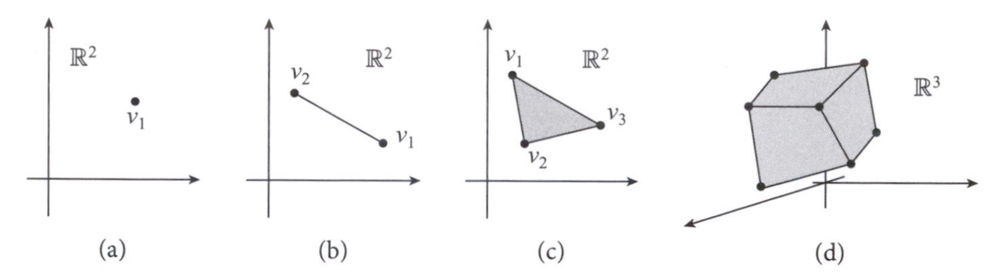

Figure 5.2 Convex hulls. (a) Convex hull in $\mathbb{R}^2$ generated by v_1; (b) Convex hull in $\mathbb{R}^2$ generated by v_1, v_2; (c) Convex hull in $\mathbb{R}^2$ generated by v_1, v_2, v_3; (d) Convex hull in $\mathbb{R}^3$.

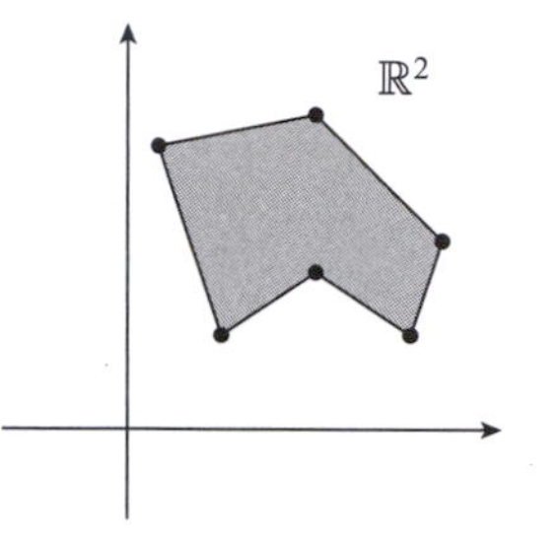

Figure 5.3 A set that is not convex.

i.e., if x_1 and x_2 belong to the set, then any point in the line segment between x_1 and x_2 also belongs to the set. Consequently, all sets in Figure 5.2 are convex, but this is not the case for the set in Figure 5.3.

5.3 Separating Hyperplane Theorem

Note. The points in the hyperplane are generally not orthogonal to v, but given any two points in the hyperplane $x_1, x_2 \in \mathcal{P}$, v it is always orthogonal to the difference $x_1 - x_2$.

An *hyperplane in* $\mathbb{R}^n$ is a set of the form

$$\mathcal{P} := \{x \in \mathbb{R}^n : v'(x - x_0) = 0\},$$

where $x_0 \in \mathbb{R}^n$ is a point that belongs to the hyperplane and $v \in \mathbb{R}^n$ a vector that is called the *normal to the hyperplane*. An *(open) half-space in* $\mathbb{R}^n$ is a set of the form

$$\mathcal{H} := \{x \in \mathbb{R}^n : v'(x - x_0) > 0\},$$

Note. Any point in the half-space can be obtained by adding to a point in the boundary a positively scaled version of the inwards-point normal.

where $x_0 \in \mathbb{R}^n$ is a point in the boundary of $\mathcal{H}$ and v is the *inwards-pointing normal* to the half-space. Each hyperplane partitions the whole space $\mathbb{R}^n$ into two half-spaces with symmetric normals.

The Separating Hyperplane Theorem is one of the key results in convex analysis. This theorem, which is illustrated in Figure 5.4, can be stated as follows:

Theorem 5.2 (Separating Hyperplane). For every convex set $\mathcal{C}$ and a point x_0 not in $\mathcal{C}$, there exists an hyperplane $\mathcal{P}$ that contains x_0 but does not intersect $\mathcal{C}$. Consequently, the set $\mathcal{C}$ is fully contained in one of the half spaces defined by $\mathcal{P}$. □

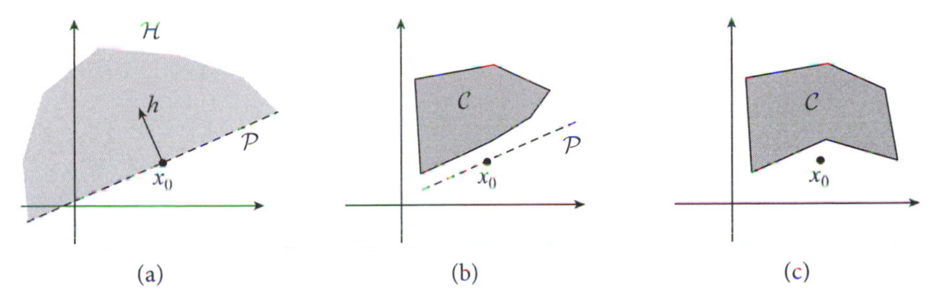

Figure 5.4 Separating Hyperplane Theorem. (a) Hyperplane and half-space; (b) Theorem correctly applied to a convex set; (c) Theorem fails for a non-convex set.

5.4 On the Way to Prove the Minimax Theorem

As mentioned before, we need to prove that for any number c, we either have

$$\underline{V}_m(A) := \max_{z\in\mathcal{Z}} \min_{y\in\mathcal{Y}} y'Az \geq c \quad \text{or} \quad \bar{V}_m(A) := \min_{y\in\mathcal{Y}} \max_{z\in\mathcal{Z}} y'Az \leq c.$$

To prove this, it will suffice to show that either there exists a $z^* \in \mathcal{Z}$ such that

$$y'Az^* \geq c,\ \forall y \in \mathcal{Y} \Rightarrow \min_{y\in\mathcal{Y}} y'Az^* \geq c \Rightarrow \underline{V}_m(A) := \max_{z\in\mathcal{Z}} \min_{y\in\mathcal{Y}} y'Az \geq c$$

or there exists a $y^* \in \mathcal{Y}$ such that

$$y^{*\prime}Az \leq c,\ \forall z \in \mathcal{Z} \Rightarrow \max_{z\in\mathcal{Z}} y^{*\prime}Az \leq c \Rightarrow \bar{V}_m(A) := \min_{y\in\mathcal{Y}} \max_{z\in\mathcal{Z}} y'Az \leq c.$$

The next result, known as the Theorem of the Alternative for Matrices, proves exactly this for the special case $c = 0$. We will later see that this can be easily generalized to an arbitrary c by appropriately redefining the matrix.

Theorem 5.3 (Theorem of the Alternative for Matrices [11]). For every $m \times n$ matrix M, one of the following statements must necessarily hold:

1. either there exists some $y^* \in \mathcal{Y}$ such that $y^{*\prime}Mz \leq 0, \forall z \in \mathcal{Z}$;
2. or there exists some $z^* \in \mathcal{Z}$ such that $y'Mz^* \geq 0, \forall y \in \mathcal{Y}$. □

Note. We can regard y^* as a policy for P_1 that guarantees an outcome no larger than zero (since it guarantees $\bar{V}_m(A) \leq 0$) and z^* as a policy for P_2 that guarantees an outcome no smaller than zero (since it guarantees $\underline{V}_m(A) \geq 0$.)

Proof of Theorem 5.3. We consider separately the cases of whether or not the vector 0 belongs to the convex hull $\mathcal{C}$ of the columns of $[\,-M_{m\times n} \quad I_m\,]$, where I_m denotes the identity matrix in $\mathbb{R}^m$ (see Figure 5.5).

Suppose first that 0 does belong to the convex hull $\mathcal{C}$ of the columns of $[-M_{m\times n} \quad I_m]$ and therefore that there exist scalars $\bar{z}_j, \bar{\eta}_j$ such that

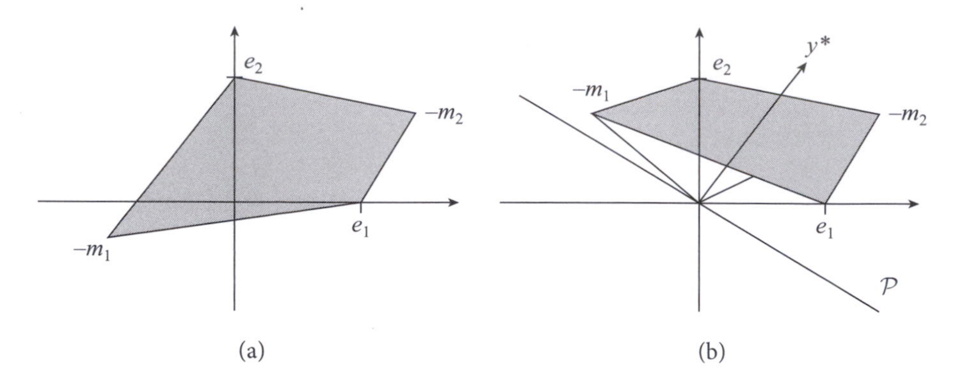

Figure 5.5 Two cases considered in the proof of Theorems 5.3 for $m = n = 2$: e_1, e_2 denote the vectors of the canonical basis of $\mathbb{R}^2$ and m_1, m_2 the columns of M. (a) 0 in the convex hull $\mathcal{C}$; (b) 0 not in the convex hull $\mathcal{C}$, with separating hyperplane $\mathcal{P}$ and inner-pointing normal y^*.

$$[\,-M_{m\times n} \quad I_m\,]\begin{bmatrix}\bar{z}_1\\ \vdots \\ \bar{z}_n \\ \bar{\eta}_1 \\ \vdots \\ \bar{\eta}_m\end{bmatrix} = 0, \quad \bar{z}_j \geq 0,\ \bar{\eta}_j \geq 0,\ \sum_j \bar{z}_j + \sum_j \bar{\eta}_j = 1. \quad (5.1)$$

First note that $\sum_j \bar{z}_j \neq 0$ since otherwise all the z_j would have to be exactly equal to zero and then so would all the $\bar{n}_j$, because of the left-hand side equality in (5.1). Defining

$$z^* := \frac{1}{\sum_j \bar{z}_j}\begin{bmatrix}\bar{z}_1\\ \vdots \\ \bar{z}_n\end{bmatrix}, \qquad \eta^* := \frac{1}{\sum_j \bar{z}_j}\begin{bmatrix}\bar{\eta}_1\\ \vdots \\ \bar{\eta}_m\end{bmatrix}$$

we conclude that $z^* \in \mathcal{Z}$ and $Mz^* = \eta^*$. Then, for every $y \in \mathcal{Y}$,

Note. Recall that this case corresponds to a game for which P_2 can guarantee an outcome no smaller than zero using the policy z^*.

$$y'Mz^* = y'\eta^* \geq 0,$$

which shows that Statement 2 holds (recall that all entries of y and η^* are non negative).

Suppose now that the 0 vector does not belong to the convex hull $\mathcal{C}$ of the columns of $[\,-M \quad I_m\,]$. In this case, we use the Separating Hyperplane Theorem to conclude that there must exist an half space $\mathcal{H}$ with 0 in its boundary that fully contains $\mathcal{C}$. Denoting by y^* the inwards-pointing normal to $\mathcal{H}$, we have that

$$\mathcal{H} = \{x \in \mathbb{R}^m : y^{*\prime}x > 0\} \supset \mathcal{C}.$$

Therefore, for every x in the convex hull $\mathcal{C}$ of the columns of $[\,-M \quad I_m\,]$, we have

$$y^*x > 0.$$

We therefore conclude that for every $\bar{z}_j \geq 0, \bar{\eta}_j \geq 0, \sum_j \bar{z}_j + \sum_j \bar{\eta}_j = 1$,

$$y^* \begin{bmatrix} -M & I_m \end{bmatrix} \begin{bmatrix} \bar{z}_1 \\ \vdots \\ \bar{z}_n \\ \bar{\eta}_1 \\ \vdots \\ \bar{\eta}_m \end{bmatrix} > 0.$$

In particular, for convex combinations with all the $\eta_j = 0$, we obtain

$$y^* M \bar{z} < 0, \quad \forall \bar{z} \in \mathcal{Z}. \tag{5.2}$$

On the other hand, from the convex combinations with $\eta_j = 1$ and all other coefficients equal to zero, we conclude that

$$y_j^* > 0, \quad \forall\, j.$$

Note. Recall that this case corresponds to a game for which P_1 can guarantee an outcome no larger than zero using the policy y^*.

In case $\sum_j y_j^* = 1$, then $y^* \in \mathcal{Y}$ (which is the hyperplane normal) provides the desired vector y^* for Statement 1. Otherwise, we simply need to rescale the normal by a positive constant to get $\sum_j y_j^* = 1$. Note that rescaling by a positive constant does not change the validity of (5.2). ■

5.5 Proof of the Minimax Theorem

With all the hard work done, we can now prove the Minimax Theorem. To do this, we pick an arbitrary constant c and use Theorem 5.3 to show that either there exists a $z^* \in \mathcal{Z}$ such that

$$y'Az^* \geq c, \quad \forall y \in \mathcal{Y} \Rightarrow \min_{y \in \mathcal{Y}} y'Az^* \geq c \Rightarrow \underline{V}_m(A) := \max_{z \in \mathcal{Z}} \min_{y \in \mathcal{Y}} y'Az \geq c \tag{5.3}$$

or there exists a $y^* \in \mathcal{Y}$ such that

$$y^{*\prime}Az \leq c, \quad \forall z \in \mathcal{Z} \Rightarrow \max_{z \in \mathcal{Z}} y^{*\prime}Az \leq c \Rightarrow \bar{V}_m(A) := \min_{y \in \mathcal{Y}} \max_{z \in \mathcal{Z}} y'Az \leq c. \tag{5.4}$$

This will be achieved by applying Theorem 5.3 to the matrix $M = A - c\mathbf{1}$, where $\mathbf{1}$ denotes a $m \times n$ matrix with all entries equal to 1. The matrix $\mathbf{1}$ has the interesting property that $y'\mathbf{1}z = 1$ for every $y \in \mathcal{Y}$ and $z \in \mathcal{Z}$. When Statement 1 in Theorem 5.3 holds, there exists some $y^* \in \mathcal{Y}$ such that

$$y^{*\prime}(A - c\mathbf{1})z = y^{*\prime}Az - c \leq 0, \quad \forall z \in \mathcal{Z}.$$

and therefore (5.4) holds. Alternatively, when Statement 2 in Theorem 5.3 holds, there exists some $z^* \in \mathcal{Z}$ such that

$$y'(A - c\mathbf{1})z^* = y'Az^* - c \geq 0, \quad \forall y \in \mathcal{Y}$$

and therefore (5.3) holds.

If the Minimax Theorem did not hold, we could then pick c such that

$$\max_{z\in\mathcal{Z}} \min_{y\in\mathcal{Y}} y'Az < c < \min_{y\in\mathcal{Y}} \max_{z\in\mathcal{Z}} y'Az,$$

which would contradict (5.3)–(5.4). Therefore there must not be a gap between the $\max_z \min_y$ and the $\min_y \max_z$.

5.6 Consequences of the minimax theorem

When we combine Theorem 4.1 with the Minimax Theorem 5.1 we conclude the following:

Notation. Since the two security levels are always equal, one just writes $V_m(A)$, which is simply called the *(mixed) value of the game.*

Corollary 5.1. Consider a game defined by a matrix A:

P5.1 A mixed saddle-point equilibrium always exist and

$$\underline{V}_m(A) := \max_{z\in\mathcal{Z}} \min_{y\in\mathcal{Y}} y'Az = \min_{y\in\mathcal{Y}} \max_{z\in\mathcal{Z}} y'Az =: \bar{V}_m(A). \tag{5.5}$$

P5.2 If y^* and z^* are mixed security policies for P_1 and P_2, respectively then (y^*, z^*) is a mixed saddle-point equilibrium and its value $y^{*\prime}Az^*$ is equal to (5.5).

P5.3 If (y^*, z^*) is a mixed saddle-point equilibrium then y^* and z^* are mixed security policies for P_1 and P_2, respectively and (5.5) is equal to the mixed saddle-point value $y^{*\prime}Az^*$.

Note. The proof of this fact follows precisely the one of Proposition 3.2.

P5.4 If (y_1^*, z_1^*) and (y_2^*, z_2^*) are mixed saddle-point equilibria then (y_1^*, z_2^*) and (y_2^*, z_1^*) are also mixed saddle-point equilibria and

$$y_1^{*\prime}Az_1^* = y_2^{*\prime}Az_2^* = y_1^{*\prime}Az_2^* = y_2^{*\prime}Az_1^*.$$

□

5.7 Practice Exercise

5.1 (Symmetric games). A game defined by a matrix A is called *symmetric* if A is skew symmetric, i.e., if $A' = -A$. For such games, show that the following statements hold:

1. $V_m(A) = 0$
2. If y^* is a mixed security policy for P_1, then y^* is also a security policy for P_2 and vice-versa.
3. If (y^*, z^*) is a mixed saddle-point equilibrium then (z^*, y^*) is also a mixed saddle-point equilibrium.

Hint: Make use of the facts that

$$\max_x f(x) = -\min_x(-f(x)), \qquad \min_w \max_x f(x) = -\max_w \min_x(-f(x)).$$

Note that the Rock-Paper-Scissors game is symmetric.

Solution to Exercise 5.1.

1. Denoting by y^* a mixed security policy for P_1, we have that

$$V_m(A) := \min_y \max_z y'Az = \max_z y^{*\prime}Az. \tag{5.6}$$

Since $y'Az$ is a scalar and $A' = -A$, we conclude that

$$y'Az = (y'Az)' = z'A'y = -z'Ay, \qquad y^{*\prime}Az = \cdots = -z'Ay^*.$$

Using this in (5.6), we conclude that

$$V_m(A) = \min_y \max_z(-z'Ay) = \max_z(-z'Ay^*).$$

We now use the "hint" to obtain

$$V_m(A) = -\max_y \min_z z'Ay = -\min_z z'Ay^*. \tag{5.7}$$

However, in mixed games $\max_y \min_z z'Ay$ is also equal to $V_m(A)$ and therefore we have

$$V_m(A) = -V_m(A) \Rightarrow V_m(A) = 0.$$

2. On the other hand, since we just saw in (5.7) that

$$\max_y \min_z z'Ay = \min_z z'Ay^*,$$

we have that y^* is indeed a mixed security policy for P_2. To prove the converse, we follow a similar reasoning starting from (5.6), but with $\max_y \min_z$ instead of $\min_z \max_y$. This results in the proof that if y^* is a mixed security policy for P_2 then it is also a mixed security policy for P_1.

3. If (y^*, z^*) is a mixed saddle-point equilibrium then both y^* and z^* are mixed security policies for P_1 and P_2, respectively. However, from the previous results we conclude that these are also security policies for P_2 and P_1, respectively, which means that (z^*, y^*) is indeed a mixed saddle-point equilibrium.

LECTURE 6

Computation of Mixed Saddle-Point Equilibrium Policies

This lecture addresses the computation of mixed saddle-point equilibria.

6.1 Graphical Method

In view of the Minimax Theorem, to find mixed saddle-point equilibria, all we need to do is to compute mixed security policies for both players (cf. Corollary 5.1).

For 2×2 games, one can compute mixed security policies in closed form using the "graphical method." To illustrate this method consider

$$A = \underbrace{\left[\begin{matrix} 3 & 0 \\ -1 & 1 \end{matrix}\right]}_{\mathsf{P}_2 \text{ choices}} \Big\} \mathsf{P}_1 \text{ choices.}$$

To compute the mixed security policy for P_1, we compute

$$\min_{y=[y_1\ y_2]} \max_{z=[z_1\ z_2]} y'Az = \min_y \max_z [\, y_1 \quad y_2 \,] \begin{bmatrix} 3 & 0 \\ -1 & 1 \end{bmatrix} \begin{bmatrix} z_1 \\ z_2 \end{bmatrix}$$

$$= \min_y \max_z z_1(3y_1 - y_2) + z_2 y_2 = \min_y \max\{3y_1 - y_2, y_2\}$$

where we used Lemma 4.1. Since $y_1 + y_2 = 1$, we must have $y_2 = 1 - y_1$ and therefore

$$\min_y \max_z y'Az = \min_{y_1 \in [0,1]} \max\{4y_1 - 1, 1 - y_1\}.$$

As shown in Figure 6.1, to find the optimal value for y_1, we draw the two lines $4y_1 - 1$ and $1 - y_1$ in the same axis, pick the maximum point-wise, and then select the point y_1^*

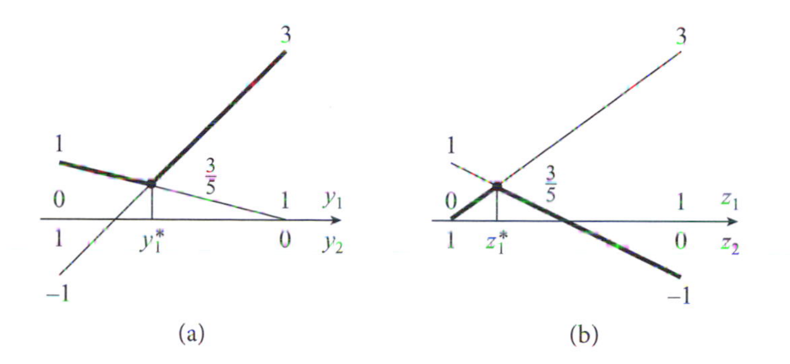

Figure 6.1 Graphical computation of mixed security policies. (a) Mixed security policy for P_1; (b) Mixed security policy for P_2.

for which the maximum is smallest. This point is the security policy and the maximum is the value of the game. As we can see in Figure 6.1(a), this results in

$$\min_y \max_z y'Az = \frac{3}{5}, \quad y_1^* = \frac{2}{5} \quad \Rightarrow \quad y^* = [\,\tfrac{2}{5} \quad \tfrac{3}{5}\,]'.$$

Similarly, to compute the mixed security policy for P_2, we compute

$$\begin{aligned}\max_{z=[z_1\ z_2]} \min_{y=[y_1\ y_2]} y'Az &= \max_z \min_y [\, y_1 \quad y_2 \,] \begin{bmatrix} 3 & 0 \\ -1 & 1 \end{bmatrix} \begin{bmatrix} z_1 \\ z_2 \end{bmatrix} \\ &= \max_z \min_y y_1(3z_1) + y_2(-z_1 + z_2) \\ &= \max_z \min\{3z_1, -z_1 + z_2\} = \max_z \min\{3z_1, 1 - 2z_1\}.\end{aligned}$$

Attention! Both security policies are unique, which means that the mixed saddle-point equilibrium is unique.

How could one get non-unique mixed saddle-point equilibria?

As we can see in Figure 6.1(b), this results in

$$\max_z \min_y y'Az = \frac{3}{5}, \quad z_1^* = \frac{1}{5} \quad \Rightarrow \quad z^* = [\,\tfrac{1}{5} \quad \tfrac{4}{5}\,]'.$$

6.2 Linear Program Solution

We now consider a systematic numerical procedure to find mixed saddle-point equilibria. Our goal is to compute

$$V_m(A) = \min_{y \in \mathcal{Y}} \max_{z \in \mathcal{Z}} y'Az = \max_{z \in \mathcal{Z}} \min_{y \in \mathcal{Y}} y'Az \tag{6.1}$$

where

$$\mathcal{Y} := \Big\{ y \in \mathbb{R}^m : \sum_i y_i = 1, y_i \geq 0, \forall i \Big\}, \quad \mathcal{Z} := \Big\{ z \in \mathbb{R}^n : \sum_j z_j = 1, z_j \geq 0, \forall j \Big\}.$$

We start by computing the "inner max" in the $\min_y \max_z$ optimization in (6.1): For a fixed $y \in \mathcal{Y}$, we have that

$$\max_{z\in\mathcal{Z}} y'Az = \max_{z\in\mathcal{Z}} \sum_{ij} y_i a_{ij} z_j = \max_{z\in\mathcal{Z}} \sum_j (z_j \sum_i y_i a_{ij}) = \max_j \sum_i y_i a_{ij}, \tag{6.2}$$

where we used Lemma 4.1. We now use a convenient equality to convert a maximization into a constrained minimization: Given a set of numbers $x_1, x_2, \ldots, x_m$,

Notation. This is often stated as "the maximum of a set of numbers is the smallest upper bound for all the numbers."

$$\max_j x_j = \min\{v \in \mathbb{R} : v \geq x_j, \forall j\}.$$

Using this in (6.2), we conclude that

$$\max_{z\in\mathcal{Z}} y'Az = \min\Big\{v : v \geq \underbrace{\sum_i y_i a_{ij}}_{\substack{j\text{th entry of row vector } y'A, \text{ or} \\ j\text{th entry of column vector } A'y}}, \forall j\Big\}.$$

Denoting by $\mathbf{1}$ a column vector consisting entirely of ones, we can re-write the condition in the set above as

Notation. Given two vectors $a, b \in \mathbb{R}^m$, we write $a \stackrel{\cdot}{\geq} b$ to mean that every entry of a is larger than or equal to the corresponding entry of b. Similar notation will be used for $\stackrel{\cdot}{\leq}$, $\stackrel{\cdot}{>}$, and $\stackrel{\cdot}{<}$.

$$v\mathbf{1} \stackrel{\cdot}{\geq} \begin{bmatrix} \sum_i y_i a_{i1} \\ \sum_i y_i a_{i2} \\ \vdots \\ \sum_i y_i a_{in} \end{bmatrix} = A'\, y.$$

This allows us to re-write (6.1) as

Notation. An optimization such as (6.3)—in which one minimizes a linear function subject to affine equality and inequality constraints—is called a *linear program.*

MATLAB® Hint 2 Linear programs can be solved numerically with matlab using `linprog` from the Optimization toolbox or the freeware Disciplined Convex Programming toolbox, also known as CVX. ▷ p. 63

Note. We shall see in Exercise 6.3 that the set of mixed policies is always convex. In particular, this means that the set of security policies can never consist of a few isolated points. ▷ p. 70

$$V_m(A) = \min_{y\in\mathcal{Y}} \min\left\{v : v\mathbf{1} \stackrel{\cdot}{\geq} A'y\right\}$$

$$= \underbrace{\begin{array}{ll} \text{minimum} & v \\ \text{subject to} & \left.\begin{array}{c} y \stackrel{\cdot}{\geq} 0 \\ \mathbf{1}y = 1 \end{array}\right\} y \in \mathcal{Y} \\ & A'y \stackrel{\cdot}{\leq} v\mathbf{1} \end{array}}_{\text{optimization over } m+1 \text{ parameters } (v, y_1, y_2, \ldots, y_m)}. \tag{6.3}$$

Solving this optimization, we directly obtain the value of the game v^* and a mixed security policy y^* for P_1. Moreover, since the security policies are precisely those that achieve the minimum in (6.3), once we have the value of the game we can obtain the set of all mixed security policies using

$$\left\{y \in \mathbb{R}^m : y \stackrel{\cdot}{\geq} 0, \mathbf{1}'y = 1, v^*\mathbf{1} \stackrel{\cdot}{\geq} A'y\right\}. \tag{6.4}$$

Alternatively, by focusing on the $\max_z \min_y$ optimization, we conclude that

$$\min_{y\in\mathcal{Y}} y'Az = \cdots = \max\{v : v\mathbf{1} \stackrel{\cdot}{\leq} Az\}$$

and therefore

$$V_m(A) = \underbrace{\begin{array}{ll} \text{minimum} & v \\ \text{subject to} & \left.\begin{array}{r} z \dot{\geq} 0 \\ \mathbf{1}z = 1 \end{array}\right\} z \in \mathcal{Z} \\ & Az \dot{\geq} v\mathbf{1} \end{array}}_{\text{optimization over } n+1 \text{ parameters } (v, z_1, z_2, \dots, z_n)} .$$

Solving this optimization, we directly obtain the value of the game v^* and a mixed security policy z^* for P_2. Moreover, once we have the value of the game, the set of all mixed security policies is given by

$$\left\{z \in \mathbb{R}^m : z \dot{\geq} 0,\, \mathbf{1}'z = 1,\, v^*\mathbf{1} \dot{\leq} Az\right\} . \tag{6.5}$$

6.3 Linear Programs with MATLAB®

MATLAB® Hint 2 (Linear programs). The command

```
[x,val]=linprog(c,Ain,bin,Aeq,beq,low,high)
```

from MATLAB®'s Optimization Toolbox numerically solves linear programs of the form

$$\begin{array}{ll} \text{minimum} & \mathtt{c}'x \\ \text{subject to} & \mathtt{Ainx} \dot{\leq} \mathtt{bin} \\ & \mathtt{Aeqx} = \mathtt{beq} \\ & \mathtt{low} \dot{\leq} x \dot{\leq} \mathtt{high} \end{array}$$

and returns the value `val` of the minimum and a vector `x` that achieves the minimum.

The vector `low` can have some or all entries equal to `-Inf` and the vector `high` have some or all entries equal to `Inf` to avoid the corresponding inequality constraints.

The same optimization could be performed with the Disciplined Convex Programming toolbox also known as CVX [4] using the following code:

Note. CVX is very convenient because its syntax is especially intuitive.

```
cvx_begin
  variables x(size(Ain,2))
  minimize c*x
  subject to
    Ain*x <= bin;
    Aeq*x  = beq;
    x     >= low;
    x     <= high;
cvx_end
```

The following CVX code can be used to find the value of a game defined by a matrix `A` and the mixed value and security policy for P_1:

```
cvx_begin
  variables v y(size(A,1))
  minimize v
  subject to
    y>=0;
    sum(y)==1;
    A'*y<=v;
cvx_end
```

whereas the following code finds the mixed value and security policy for P_2:

```
cvx_begin
  variables v z(size(A,2))
  maximize v
  subject to
    z>=0;
    sum(z)==1;
    A*z>=v;
cvx_end
```

□

6.4 Strictly Dominating Policies

Consider a game specified by an $m \times n$ matrix A, with m actions for P_1 and n actions for P_2:

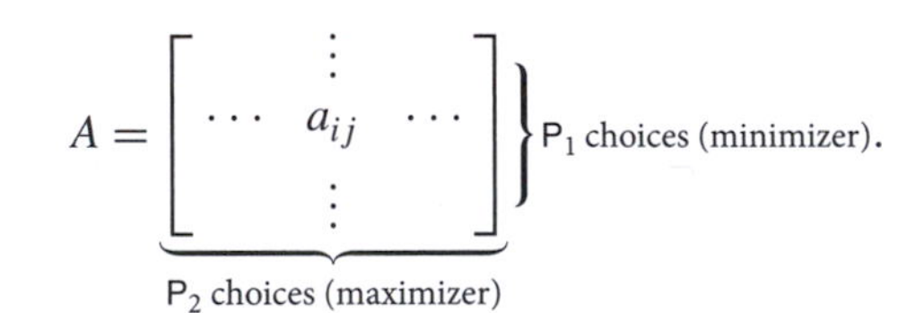

We say that *row i strictly dominates row k* if

$$a_{ij} < a_{kj} \quad \forall j,$$

which means that no matter what P_2 does, the minimizer P_1 is always better off by selecting row i instead of row k. In practice, this means that

- in pure policies, P_1 will never select row k
- in mixed policies, P_1 will always select row k with probability zero, i.e., $y_k^* = 0$ for any security/saddle-point policy.

Conversely, we say that *column j strictly dominates column ℓ* if

$$a_{ij} > a_{i\ell} \quad \forall i,$$

which means that no matter what P_1 does, the maximizer P_2 is always better off by selecting column j instead of column ℓ. In practice, this means that

- in pure policies, P_2 will never select column ℓ
- in mixed policies, P_2 will always select column ℓ with probability zero, i.e., $z^*_\ell = 0$ for any security/saddle-point policy.

Finding dominating rows/columns allows one to reduce the size of the problem that needs to be solved, as we can:

1. remove from A any rows/columns that are strictly dominated
2. compute (pure or mixed) saddle-point equilibria for the smaller game
3. recover the saddle-point equilibria for the original:
   - for pure policies the saddle-point equilibria are the same, modulo some re-indexing to account for the fact that indexes of the rows/columns may have changed
   - for mixed policies one may need to insert zero entries corresponding to the columns/rows that were removed.

By removing strictly dominated rows/columns one cannot lose security policies so all security policies for the original (larger) game correspond to security policies for the reduced game.

Example 6.1 (Strictly dominating policies). The following matrix defines a zero-sum matrix game for which the minimizer has 3 actions and the maximizer has 4 actions:

$$A = \underbrace{\begin{bmatrix} 3 & -1 & 0 & -1 \\ 4 & 1 & 2 & 0 \\ -1 & 0 & 1 & 0 \end{bmatrix}}_{\mathsf{P}_2 \text{ choices}} \left.\vphantom{\begin{bmatrix} 3 \\ 4 \\ -1 \end{bmatrix}}\right\} \mathsf{P}_1 \text{ choices}.$$

Since the 2nd row is strictly dominated by the 1st row, we can reduce the matrix to

$$A^\dagger = \begin{bmatrix} 3 & -1 & 0 & -1 \\ -1 & 0 & 1 & 0 \end{bmatrix}.$$

We now observe that both the 2nd and 4th column are (strictly) dominated by the 3rd column, therefore we can further reduce the matrix to

$$A^\ddagger = \begin{bmatrix} 3 & 0 \\ -1 & 1 \end{bmatrix},$$

for which we found the following value and mixed security/saddle-point equilibrium policies in Section 6.1:

$$V(A^\ddagger) = \frac{3}{5}, \qquad y^* = \begin{bmatrix} \frac{2}{5} \\ \frac{3}{5} \end{bmatrix}, \qquad z^* = \begin{bmatrix} \frac{1}{5} \\ \frac{4}{5} \end{bmatrix}.$$

We thus conclude that the original game has the following value and mixed security/saddle-point equilibrium policies:

$$V(A) = \frac{3}{5}, \qquad y^* = \begin{bmatrix} \frac{2}{5} \\ 0 \\ \frac{3}{5} \end{bmatrix}, \qquad z^* = \begin{bmatrix} \frac{1}{5} \\ 0 \\ \frac{4}{5} \\ 0 \end{bmatrix}. \qquad \square$$

6.5 "WEAKLY" DOMINATING POLICIES

Notation. One generally simply says that *row i dominates row k* when referring to weakly dominating policies.

Note. If all inequalities hold with equality, the two actions are actually the same, which does not make much sense.

Note. Although now, the minimizer may gain nothing from this, if the maximizer ends up selecting a column for which $a_{ij} = a_{kj}$.

We say that *row i (weakly) dominates row k* if

$$a_{ij} \leq a_{kj} \quad \forall j,$$

which means that no matter what P_2 does, the minimizer P_1 loses nothing by selecting row i instead of row k. Conversely, we say that *column j (weakly) dominates column ℓ* if

$$a_{ij} \geq a_{i\ell} \quad \forall i,$$

which means that no matter what P_1 does, the maximizer P_2 loses nothing by selecting column j instead of column ℓ.

To find the value and security policies/saddle-point equilibria for a game, one may still remove weakly dominated rows/columns and be sure that

- the value of the reduced game is the same as the value of the original game, and
- one can reconstruct security policies/saddle-point equilibria for the original game from those for the reduced game.

Contrary to what happened with strict domination, now one may lose some security policies/saddle-point equilibria that were available for the original game but that have no direct correspondence in the reduced game.

A game is said to be *maximally reduced* when no row or column dominates another one. Saddle-point equilibria of maximally reduced games are called *dominant*.

Example 6.2 (Weakly dominating policies). The following matrix defines a zero-sum matrix game for which the minimizer has 2 actions and the maximizer has 2 actions:

$$A = \underbrace{\begin{bmatrix} 0 & 1 \\ -1 & 1 \end{bmatrix}}_{\mathsf{P}_2 \text{ choices}} \Big\} \, \mathsf{P}_1 \text{ choices}.$$

This game has value $V(A) = 1$ and two pure saddle-point equilibria (1, 2) and (2, 2). However, this game can be reduced as follows:

Note. The (2, 2) pair of pure policies corresponds to the $\left(\begin{bmatrix} 0 \\ 1 \end{bmatrix}, \begin{bmatrix} 0 \\ 1 \end{bmatrix}\right)$ pair of mixed policies.

$$A = \begin{bmatrix} 0 & 1 \\ -1 & 1 \end{bmatrix} \xrightarrow[\text{(remove latter)}]{\text{2nd row dominates 1st}} A^\dagger := [\,-1 \quad 1\,] \xrightarrow[\text{(remove latter)}]{\text{2nd column dominates 1st}} A^\ddagger := [\,1\,],$$

from which one would obtain the pure saddle-point equilibrium (2, 2). Alternatively, this game can also be reduced as follows:

$$A = \begin{bmatrix} 0 & 1 \\ -1 & 1 \end{bmatrix} \xrightarrow[\text{(remove latter)}]{\text{2nd column dominates 1st}} A^{\dagger} := \begin{bmatrix} 1 \\ 1 \end{bmatrix} \xrightarrow[\text{(remove latter)}]{\text{1st row dominates 2nd}} A^{\ddagger} := [\,1\,],$$

from which one would obtain the pure saddle-point equilibrium (1, 2). □

Note. The (1, 2) pair of pure policies corresponds to the $\left(\begin{bmatrix}1\\0\end{bmatrix}, \begin{bmatrix}0\\1\end{bmatrix}\right)$ pair of mixed policies.

6.6 Practice Exercises

6.1 (Mixed security levels/policies—graphical method). For each of the following two zero-sum matrix games compute the average security levels and all mixed security policies for both players.

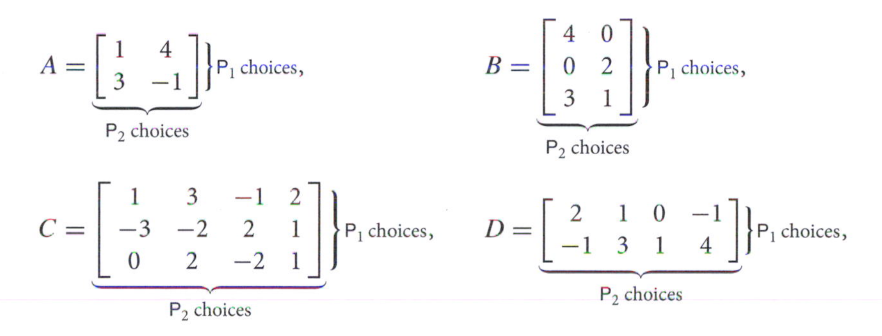

$$A = \underbrace{\begin{bmatrix} 1 & 4 \\ 3 & -1 \end{bmatrix}}_{\mathsf{P}_2 \text{ choices}} \Big\} \mathsf{P}_1 \text{ choices}, \qquad B = \underbrace{\begin{bmatrix} 4 & 0 \\ 0 & 2 \\ 3 & 1 \end{bmatrix}}_{\mathsf{P}_2 \text{ choices}} \Big\} \mathsf{P}_1 \text{ choices},$$

$$C = \underbrace{\begin{bmatrix} 1 & 3 & -1 & 2 \\ -3 & -2 & 2 & 1 \\ 0 & 2 & -2 & 1 \end{bmatrix}}_{\mathsf{P}_2 \text{ choices}} \Big\} \mathsf{P}_1 \text{ choices}, \qquad D = \underbrace{\begin{bmatrix} 2 & 1 & 0 & -1 \\ -1 & 3 & 1 & 4 \end{bmatrix}}_{\mathsf{P}_2 \text{ choices}} \Big\} \mathsf{P}_1 \text{ choices},$$

Use the graphical method.

Hint: for 3 × 2 and 2 × 3 games start by computing the average security policy for the player with only two actions.

Solution to Exercise 6.1.

1. For the matrix A

$$\begin{aligned} V_m(A) &= \min_y \max_z \; y_1 z_1 + 4 y_1 z_2 + 3 y_2 z_1 - y_2 z_2 \\ &= \min_y \max\{y_1 + 3y_2, 4y_1 - y_2\} \\ &= \min_y \max\{-2y_1 + 3, 5y_1 - 1\} = \frac{13}{7} \end{aligned}$$

with the following (unique) mixed security policy for P_1: $y^* := [\,\frac{4}{7} \;\; \frac{3}{7}\,]'$, and

$$\begin{aligned} V_m(A) &= \max_z \min_y \; y_1 z_1 + 4 y_1 z_2 + 3 y_2 z_1 - y_2 z_2 \\ &= \max_z \max\{z_1 + 4z_2, 3z_1 - z_2\} \\ &= \max_z \max\{-3z_1 + 4, 4z_1 - 1\} = \frac{13}{7} \end{aligned}$$

with the following (unique) mixed security policy for P_2: $z^* := [\,\frac{5}{7} \;\; \frac{2}{7}\,]'$. Consequently, this game has a single mixed saddle-point equilibrium $(y^*, z^*) = ([\,\frac{4}{7} \;\; \frac{3}{7}\,]', [\,\frac{5}{7} \;\; \frac{2}{7}\,]')$.

2. For the matrix B

$$\begin{aligned} V_m(B) &= \max_z \min_y 4y_1z_1 + 2y_2z_2 + 3y_3z_1 + y_3z_2 \\ &= \max_z \min\{4z_1, 2z_2, 3z_1 + z_2\} \\ &= \max_z \min\{4z_1, 2 - 2z_1, 2z_1 + 1\} = \frac{4}{3} \end{aligned}$$

with the following sole mixed security policy for P_2: $z^* := [\,\frac{1}{3} \;\; \frac{2}{3}\,]'$. Since P_1 has more than 2 actions, we cannot use the graphical method to find the mixed security policies for this player. However, from (6.4) we know that these policies must satisfy

$$\frac{4}{3}\mathbf{1} \dot{\geq} A'y = \begin{bmatrix} 4y_1 + 3y_3 \\ 2y_2 + y_3 \end{bmatrix} = \begin{bmatrix} y_1 - 3y_2 + 3 \\ -y_1 + y_2 + 1 \end{bmatrix}$$

$$\Leftrightarrow y_1 \leq -\frac{5}{3} + 3y_2,\ y_1 \geq y_2 - \frac{1}{3},\ (y_1 \leq 1 - y_2)$$

which has a single solution $y^* := [\,\frac{1}{3} \;\; \frac{2}{3} \;\; 0\,]'$. Consequently, this game has a single mixed saddle-point equilibrium $(y^*, z^*) = \big([\,\frac{1}{3} \;\; \frac{2}{3} \;\; 0\,]', [\,\frac{1}{3} \;\; \frac{2}{3}\,]'\big)$.

3. For the matrix C, row 3 strictly dominates over row 1 and column 2 strictly dominates over column 1. We can therefore reduce the game to

$$C^\dagger := \begin{bmatrix} -2 & 2 & 1 \\ 2 & -2 & 1 \end{bmatrix}.$$

For this matrix,

$$\begin{aligned} V_m(C^\dagger) &= \min_y \max\{-2y_1 + 2y_2, 2y_1 - 2y_2, y_1 + y_2\} \\ &= \min_y \max\{-4y_1 + 2, 4y_1 - 2, 1\} = 1 \end{aligned}$$

which has multiple minima for $y_1 \in [\frac{1}{4}, \frac{3}{4}]$. These correspond to the following security policies for P_1 in the original game: $y^* := [0 \;\; y_1 \;\; 1 - y_1]'$ for any $y_1 \in [\frac{1}{4}, \frac{3}{4}]$. Since P_2 has more than 2 actions even for the reduced game $C\dagger$, we cannot use the graphical method to find the mixed security policies for this player. However, from (6.5) we know that these policies must satisfy

$$\mathbf{1} \dot{\leq} Az = \begin{bmatrix} -2z_1 + 2z_2 + z_3 \\ 2z_1 - 2z_2 + z_3 \end{bmatrix} = \begin{bmatrix} -3z_1 + z_2 + 1 \\ z_1 - 3z_2 + 1 \end{bmatrix}$$

$$\Leftrightarrow z_2 \geq 3z_1,\ z_2 \leq \frac{1}{3}z_1, \quad (z_2 \leq 1 - z_1)$$

which has a single solution $z_1 = z_2 = 0$. This corresponds to the following security policy for P_2 in the original game: $z^* := [\,0 \;\; 0 \;\; 0 \;\; 1\,]'$. Consequently, this game has the following family of mixed saddle-point equilibria $(y^*, z^*) = ([\,0 \quad y_1 \quad 1-y_1\,]', [\,0 \quad 0 \quad 0 \quad 1\,]'), y_1 \in [\frac{1}{4}, \frac{3}{4}]$.

4. For the matrix D, column 2 strictly dominates column 3 and we can therefore reduce the game to

$$D^\dagger := \begin{bmatrix} 2 & 1 & -1 \\ -1 & 3 & 4 \end{bmatrix}.$$

For this matrix,

$$\begin{aligned} V_m(D^\dagger) &= \min_y \max\{2y_1 - y_2, y_1 + 3y_2, 4y_2 - y_1\} \\ &= \min_y \max\{3y_1 - 1, 3 - 2y_1, 4 - 5y_1\} = \frac{7}{5} \end{aligned}$$

which has a single minimum for $y_1 = \frac{4}{5}$. This corresponds to the following security policies for P_1 in the original game: $y^* := [\,\frac{4}{5} \quad \frac{1}{5}\,]'$. Since P_2 has more than 2 actions even for the reduced game $D\dagger$, we cannot use the graphical method to find the mixed security policies for this player. However, from (6.4) we know that these policies must satisfy

$$\frac{7}{5}\mathbf{1} \dot{\leq} Az = \begin{bmatrix} 2z_1 + z_2 - z_3 \\ -z_1 + 3z_2 + 4z_3 \end{bmatrix} = \begin{bmatrix} 3z_1 + 2z_2 - 1 \\ -5z_1 - z_2 + 4 \end{bmatrix}$$

$$\Leftrightarrow z_2 \geq -\frac{3}{2}z_1 + \frac{6}{5}, z_2 \leq -5z_1 + \frac{13}{5}, \quad (z_2 \leq 1 - z_1)$$

which has a single solution $z_1 = \frac{2}{5}$, $z_2 = \frac{3}{5}$. This corresponds to the following security policy for P_2 in the original game: $z^* := [\,\frac{2}{5} \quad \frac{3}{5} \quad 0 \quad 0\,]'$. Consequently, this game has the unique mixed saddle-point equilibrium $(y^*, z^*) = ([\,\frac{4}{5} \quad \frac{1}{5}\,]', [\,\frac{2}{5} \quad \frac{3}{5} \quad 0 \quad 0\,]')$.

6.2 (Mixed security levels/policies—LP method). For each of the following two zero-sum matrix games compute the average security levels and a mixed security policy

$$A = \underbrace{\begin{bmatrix} 3 & 1 \\ 2 & 2 \\ 1 & 3 \end{bmatrix}}_{\mathsf{P}_2 \text{ choices}} \Big\} \mathsf{P}_1 \text{ choices}. \qquad B = \underbrace{\begin{bmatrix} 0 & 1 & 2 & 3 \\ 1 & 0 & 1 & 2 \\ 0 & 1 & 0 & 1 \\ -1 & 0 & 1 & 0 \end{bmatrix}}_{\mathsf{P}_2 \text{ choices}} \Big\} \mathsf{P}_1 \text{ choices}.$$

Solve this problem numerically using MATLAB®.

Solution to Exercise 6.2.

1. For the matrix A we use the following CVX code to compute the mixed value and the security policies for P_2 and P_1:

```
A=[3,1;2,2;1,3];

cvx_begin;
  variables v z(size(A,2));
  maximize v;
  subject to
    z>=0;
    sum(z)==1;
    A*z>=v;
cvx_end;

cvx_begin;
  variables v y(size(A,1));
  minimize v;
  subject to
    y>=0;
    sum(y)==1;
    A'*y<=v;
cvx_end;
```

which resulted in the mixed value $V_m(A) = 2$ and a saddle-point

$$(y^*, z^*) = \left([\, \tfrac{1}{3} \quad \tfrac{1}{3} \quad \tfrac{1}{3} \,]', [\, \tfrac{1}{2} \quad \tfrac{1}{2} \,]' \right).$$

However, this matrix has multiple saddle-point equilibria and for y^* so you may get any distribution of the form $[\, \lambda \quad 1 - 2\lambda \quad \lambda \,]', \forall \lambda \in [0, 1/2]$.

2. For the matrix B we use the same CVX but with the first line replaced by

```
A=[0,1,2,3;1,0,1,2;0,1,0,1;-1,0,1,0];
```

which results in the mixed value $V_m(B) = \frac{1}{2}$ and a saddle-point

$$(y^*, z^*) = \left([\, 0 \quad 0 \quad \tfrac{1}{2} \quad \tfrac{1}{2} \,]', [\, 0 \quad .1858 \quad \tfrac{1}{2} \quad .3142 \,]' \right).$$

However, this matrix has multiple saddle-point equilibria for z^* so you may get any distribution of the form $[\, 0 \quad \frac{\lambda}{2} \quad \frac{1}{2} \quad \frac{1-\lambda}{2} \,]', \forall \lambda \in [0, 1]$.

6.7 Additional Exercise

6.3 (Set of security policies). Verify that the set of mixed security policies for a player is always convex.

LECTURE 7

Games in Extensive Form

This lecture introduces games in extensive form and defines several key concepts for such games.

7.1 Motivation

The following matrix defines a zero-sum matrix game for which the minimizer has 2 actions and the maximizer has 3 actions:

$$A = \underbrace{\begin{bmatrix} 1 & 3 & 0 \\ 6 & 2 & 7 \end{bmatrix}}_{\substack{\text{P}_2 \text{ choices} \\ \text{(Left, Middle, Right)}}} \Big\} \substack{\text{P}_1 \text{ choices} \\ \text{(Top,Bottom)}} . \tag{7.1}$$

For this game we have

$$\underline{V}(A) = \max_j \min_i a_{ij} = 2$$

$$\bar{V}(A) = \min_i \max_j a_{ij} = 3$$

$$V_m(A) = \min_y \max_z y'Az = \max_z \min_y y'Az = \frac{8}{3} \approx 2.667.$$

Assuming rational players, each of these values (and the corresponding policies) are particularly meaningful for a particular information structure of the game:

- $\underline{V}(A)$ will be the outcome of the game when the maximizer P_2 plays first (and the minimizer P_1 knows P_2's choice before selecting an action).
- $\bar{V}(A)$ will be the outcome of the game when the minimizer P_1 plays first (and the maximizer P_2 knows P_1's choice before selecting an action).
- $V_m(A)$ will be the expected value of the outcome when both players play simultaneously (none knowing the others choice before selecting their actions).

However, the matrix description, by itself, does not capture the information structure of the game and, in fact, *other information structures are possible.*

7.2 Extensive Form Representation

Note. Actually, one can imagine games represented by general directed graphs that are not decision trees.

Notation. We will generally name the information sets using Greek letters.

Exercise. Students so inclined, may think about what data structures would be most appropriate to represent games in extensive form.

An extensive form representation of a zero-sum two-person game is a decision tree.

1. Each *node* of the tree must be associated with one player (P_1 or P_2).
2. The *links* emanating from one node correspond to decisions made by the corresponding player.
3. All nodes must be enclosed in *dashed boxes*, called *information sets (IS)*. Each information set may contain one or more nodes of the same player. All nodes in the same information set are indistinguishable for the corresponding player and must have precisely the same alternatives.
4. The different *leafs* of the tree correspond to the different final *outcomes* of the game and should be labeled with the corresponding cost/reward.

Figure 7.1 shows several alternative games in extensive form, all represented by the same matrix (7.1) of outcomes. For this 2×3 matrix of outcomes there is a total of

$$\underbrace{1}_{\text{P}_1 \text{ plays first}} + \underbrace{1}_{\text{simultaneous}} + \underbrace{1}_{\substack{\text{P}_2 \text{ plays first} \\ \text{with full information}}} + \underbrace{3}_{\substack{\text{P}_2 \text{ plays first} \\ \text{with partial information}}} = 6$$

possible information structures, all shown in Figure 7.1.

7.3 Multi-Stage Games

So far we have focused on *single-stage games* in which each player must choose an action only once per game. *Multi-stage games* consist of a sequence of rounds (or stages) and in each stage the players have the opportunity to select one action.

As shown in Figure 7.2, the extensive form representation of multi-stage games is very similar to the one for single-stage games, except that now the tree can have more than two levels. In fact, the tree will generally have twice as many levels as the number of stages.

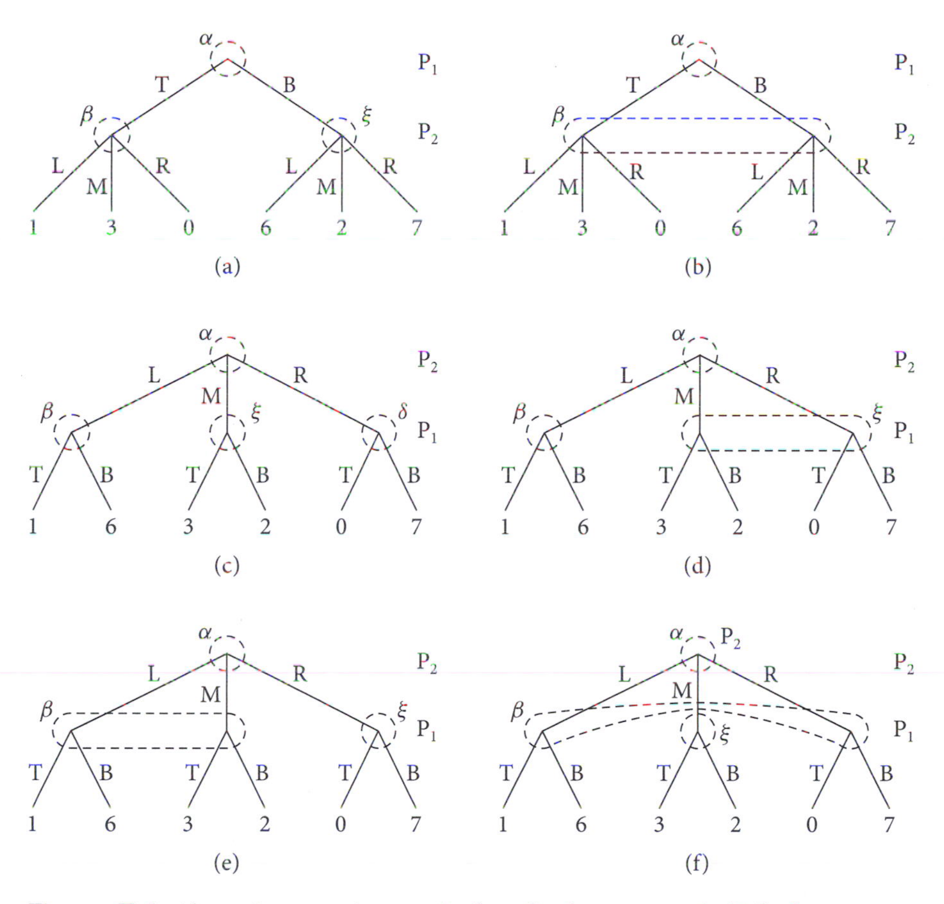

Figure 7.1 Alternative games in extensive form for the same matrix (7.1) of outcomes. (a) $P_1 - P_2$: P_1 plays first, and P_2 plays knowing P_1's decision. (b) Simultaneous: Both players play at the same time without knowing each others choice. (c) $P_2 - P_1$: P_2 plays first, and P_1 plays knowing P_2's decision. (d) Mixed: P_2 plays first, if P_2 plays L then P_1 knows about this choice, but if P_2 plays M or R, then P_1 does not know which was played. (e) Mixed: P_2 plays first, if P_2 plays R then P_1 knows about this choice, but if P_2 plays L or M, then P_1 does not know which was played. (f) Mixed: P_2 plays first, if P_2 plays M then P_1 knows about this choice, but if P_2 plays L or R, then P_1 does not know which was played.

Some multi-stage games have a *variable number of stages*, depending on the actions of the players [cf. Figure 7.2(b)]. Chess is an example of such a game, because some actions of the players will lead to the termination of the game with very few stages, whereas other sets of actions may require a large number of stages before the game is over.

The *information sets* in multi-stage games may span over different stages as long as they only contain nodes of the same player and all nodes within the set exhibit exactly the same possible actions for that player [cf. Figure 7.2(c)]. When an information set spans several stages, that particular player does not know at which stage play is taking place.

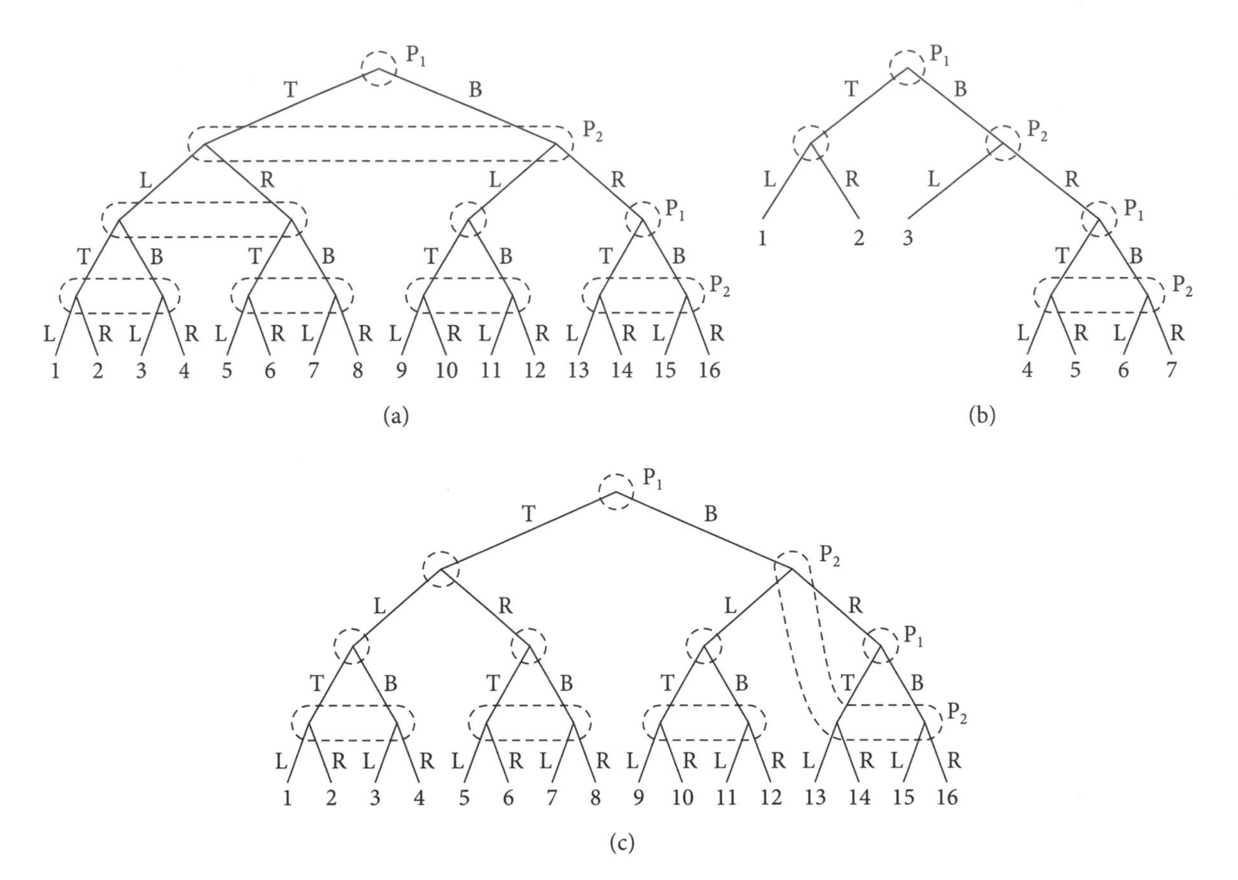

Figure 7.2 Examples of extensive form representations of multi-stage games. (a) Two-stage game with different types of information sets, depending on the players' actions. (b) Variable-stage game that may have one or two stages, depending on the actions of the players. (c) Two stage game with an information set for P_2 that spans across stages: If P_1 initially chooses B, then P_2 must make a decision without knowing the stage of play (this would be a very "memory-constrained" player).

7.4 Pure Policies and Saddle-Point Equilibria

For games in extensive form, a *pure policy* for the player P_i is a decision rule (i.e., a function) that associates one action to each information set of this player.

For example, for the game $P_1 - P_2$ in Figure 7.1, the following are possible policies π_1, π_2 for players P_1 and P_2, respectively:

$$\pi_1(\text{IS}) = \{\, \text{T} \quad \text{if IS} = \alpha \qquad \pi_2(\text{IS}) = \begin{cases} \text{M} & \text{if IS} = \beta \\ \text{R} & \text{if IS} = \xi. \end{cases}$$

Note. The total number of possible distinct pure policies grows very fast with the number of actions and information sets. In fact, it grows exponentially with the number of information sets.

The total number of distinct pure policies for a given player is given by

$$\underbrace{(\text{\# actions of 1st IS}) \times (\text{\# actions of 2nd IS}) \times \cdots \times (\text{\# actions of last IS})}_{\text{product over all information sets for that player}}.$$

So for the alternate play $\mathsf{P}_1 - \mathsf{P}_2$ game in Figure 7.1 we have a total of

- 2 pure policies for P_1, and
- $3^2 = 9$ pure policies for P_2.

Whereas for the simultaneous play game in Figure 7.1 we have a total of

- 2 pure policies for P_1, and
- 3 pure policies for P_2.

Now that we have formalized the notion of policy for a game in extensive form, we can use the concepts of security levels, security policies, and saddle-point equilibria introduced in Section 4.5 for general games, with the understanding that:

1. the *action spaces* are the sets Γ_1 and Γ_2 of all pure policies for players P_1 and P_2, respectively; and
2. for a particular pair of pure policies $\gamma \in \Gamma_1$, $\sigma \in \Gamma_2$ we denote by $J(\gamma, \sigma)$ the *outcome of the game* when player P_1 uses policy γ and P_2 uses policy σ.

Attention! With the definitions in Section 4.5 we recover all the results in Proposition 4.2 regarding the relationship between security levels/policies and saddle-point equilibria. However, now security policies always exist since the action spaces are finite.

In particular, this leads to the following definition of a saddle-point equilibrium:

Definition 7.1 (Pure saddle-point equlibrium). A pair of policies $(\gamma^*, \sigma^*) \in \Gamma_1 \times \Gamma_2$ is a *pure saddle-point equilibrium* if

$$J(\gamma^*, \sigma^*) \leq J(\gamma, \sigma^*), \quad \forall \gamma \in \Gamma_1 \tag{7.2}$$

$$J(\gamma^*, \sigma^*) \geq J(\gamma^*, \sigma), \quad \forall \sigma \in \Gamma_2 \tag{7.3}$$

and $J(\gamma^*, \sigma^*)$ is called the saddle-point value.

7.5 Matrix Form for Games in Extensive Form

Any game in *extensive form* can be converted into an equivalent game in *matrix form* by regarding each policy of the game in extensive form as a possible action in a game in matrix form.

To this effect, take a game in extensive form for which

$$\Gamma_1 = \{\gamma_1, \gamma_2, \ldots, \gamma_m\}, \qquad \Gamma_2 = \{\sigma_1, \sigma_2, \ldots, \sigma_n\}$$

are the sets of all pure policies for players P_1 and P_2, respectively. We then define an $m \times n$ matrix A_{ext}, where

Notation. The matrix A_{ext} constructed by this procedure is called the *matrix form representation* of the game in extensive form.

Note. The size of the matrix A_{ext} grows exponentially with the number of actions and information sets, which limits the use of this procedure to fairly small games.

1. each row of A_{ext} corresponds to one pure policy for P_1 and therefore m is equal to the total number of distinct pure policies in Γ_1,
2. each column of A_{ext} corresponds to one pure policy for P_2 and therefore n is equal to the total number of distinct pure policies in Γ_2,
3. the entry a_{ij} of A_{ext} is equal to the outcome of the game $J(\gamma_i, \sigma_j)$ when P_1 selects the pure policy γ_i corresponding to the ith row of A_{ext} and P_2 selects the pure policy σ_j corresponding to the jth column.

The matrix game defined by A_{ext} should be interpreted as being played under *simultaneous play*, since the players of the extensive game choose their policies before the game starts. Recall that each row/column of A_{ext} represents the choice of a policy, not an action.

With the matrix A_{ext} constructed in this way, we can re-write (7.2)–(7.3) as

$$a_{i^*j^*} \le a_{ij^*}, \quad \forall i$$
$$a_{i^*j^*} \ge a_{i^*j}, \quad \forall j,$$

where i^* is the row of A_{ext} corresponding to the policy γ^* and j^* is the column of A_{ext} corresponding to the policy σ^*. We therefore conclude that (γ^*, σ^*) is a pure saddle-point equilibrium for the game in extensive form if and only if (i^*, j^*) is a pure saddle-point equilibrium for the matrix game defined by A_{ext}.

Proposition 7.1. The game in extensive form has a pure saddle-point equilibrium if and only if the matrix game defined by A_{ext} has a pure saddle-point equilibrium, which is to say if and only if $\underline{V}(A_{\text{ext}}) = \bar{V}(A_{\text{ext}})$. Moreover, if such equilibria exist, they have the same saddle-point value. □

Example 7.1. Going back to the $\mathsf{P}_1 - \mathsf{P}_2$ game in extensive form in Figure 7.1, suppose that we enumerate the policies available for P_1 and P_2 as follows:

P_1:

IS	γ_1	γ_2
α	T	B

P_2:

IS	σ_1	σ_2	σ_3	σ_4	σ_5	σ_6	σ_7	σ_8	σ_9
β	L	L	L	M	M	M	R	R	R
ξ	L	M	R	L	M	R	L	M	R

In this case, we obtain

$$A_{\text{ext}} = \underbrace{\begin{bmatrix} 1 & 1 & 1 & 3 & 3 & 3 & 0 & 0 & 0 \\ 6 & 2 & 7 & 6 & 2 & 7 & 6 & 2 & 7 \end{bmatrix}}_{\substack{\mathsf{P}_2 \text{ policies} \\ (\sigma_1, \sigma_2, \ldots, \sigma_9)}} \Big\} \begin{matrix} \mathsf{P}_1 \text{ policies} \\ (\gamma_1, \gamma_2). \end{matrix} \tag{7.4}$$

For this matrix $\underline{V}(A_{\text{ext}}) = \bar{V}(A_{\text{ext}}) = 3$ and this matrix has two pure saddle-point equilibria: (1, 4) and (1, 6). Consequently, the $\mathsf{P}_1 - \mathsf{P}_2$ game in extensive form in Figure 7.1 also has two pure saddle-point equilibria: (γ_1, σ_4) and (γ_1, σ_6).

However, (γ_1, σ_6) is the dominating saddle-point equilibrium. Indeed,

Note 7. It was not a coincidence that this "large"matrix game reduced to a scalar game with no choices for either player . ▷ p. 77

$$A_{\text{ext}} \xrightarrow[\text{col. 3 dominates over 1 \& 2, col. 6 dominates over 4 \& 5}]{\text{col. 9 dominates over 7 \& 8}} A^{\dagger}_{\text{ext}} := \begin{bmatrix} 1 & 3 & 0 \\ 7 & 7 & 7 \end{bmatrix} \tag{7.5}$$

$$\xrightarrow[\text{row 1 dominates over 2}]{} A^{\ddagger}_{\text{ext}} := [\,1 \quad 3 \quad 0\,] \xrightarrow[\text{col. 2 dominates over 1 \& 3}]{} A^{\flat}_{\text{ext}} := [\,3\,].$$

Going back to the original game, this corresponds precisely to the pure saddle-point equilibrium (γ_1, σ_6):

$$\gamma_1(\text{IS}) = \{\, \text{T} \quad \text{if IS} = \alpha \qquad \sigma_6(\text{IS}) = \begin{cases} \text{M} & \text{if IS} = \beta \text{ (i.e., if } \mathsf{P}_1 \text{ chooses T)} \\ \text{R} & \text{if IS} = \xi \text{ (i.e., if } \mathsf{P}_1 \text{ chooses B).} \end{cases} \tag{7.6}$$

□

Note 7. It is not a coincidence that the matrix game defined by (7.4) can be reduced to a (scalar) game with no choices for either player. We shall see shortly that this always happens for certain types of information structure (cf. Corollary 7.1 in Section 7.6).

Note also that the policies in (7.6) are precisely the ones that would have been found by a direct analysis of the tree, similar to what was done in Section 2.1.2. In fact, the reduction in (7.5) is essentially a formalization of the tree analysis that we have done before.

However, this does not invalidate the fact that

$$\gamma_1(\text{IS}) = \{\, \text{T} \quad \text{if IS} = \alpha \qquad \sigma_4(\text{IS}) = \begin{cases} \text{M} & \text{if IS} = \beta \text{ (i.e., if } \mathsf{P}_1 \text{ chooses T)} \\ \text{L} & \text{if IS} = \xi \text{ (i.e., if } \mathsf{P}_1 \text{ chooses B)} \end{cases}$$

is also a (non-dominating) saddle-point equilibrium. □

7.6 Recursive Computation of Equilibria for Single-Stage Games

The computation of pure saddle-point equilibria for a game in extensive form by computing its matrix form representation A_{ext} can only be done for fairly small games, because the size of A_{ext} grows exponentially with the number of information sets.

An alternative approach that is computationally much more attractive consists of a recursive procedure that starts at the bottom of the tree and moves upwards. For *single-stage games* the basic procedure consists of the four steps described below. For simplicity, in listing these steps we assume that the root of the tree corresponds to a decision point by P_1 (i.e., P_1 is the "first-acting" player).

Example. Figure 7.3 contains examples illustrating how to apply this procedure. ▹ p. 78

1. Construct a matrix game corresponding to each information set of P_2. Each matrix game will have one action of P_1 for each edge entering the information set and one action of P_2 for each edge leaving the information set.
2. Compute pure saddle-point equilibria for each of the matrix games constructed.

   If any of these matrix games does not have a pure saddle-point equilibrium, this method fails. Otherwise, P_2's pure security/saddle-point policy for the game in extensive form should map to each information set the action corresponding to P_2's pure security policy in the corresponding matrix game.
3. Replace each information set by the value of the corresponding pure saddle-point equilibrium. The link leading to this value should be labeled with P_1's pure policy in the corresponding matrix game.
4. The player P_1 chooses the policy corresponding to the pure saddle-point value that is most favorable for P_1.

Note. If there are multiple saddle-points any one will do.

Note. Recall that a policy is a function that maps each information set with one action.

Note. At this point P_2 disappears and one is left with a single-player optimization.

Note. If multiple policies achieve the same value, any one will do.

This procedure is guaranteed to result in a pure saddle-point equilibrium, from which we conclude the following:

Proposition 7.2. For any zero-sum single-stage game in extensive form, if every matrix game corresponding to the information sets of the second-acting player has a pure saddle-point equilibrium, then the game in extensive form has a pure saddle-point equilibrium. □

Note. Proposition 7.2 is a special case of a more general result that also applies to multi-stage games, so we will postpone to Section 7.9 the discussion of why it is correct. ▹ p. 83

1. Initial games.

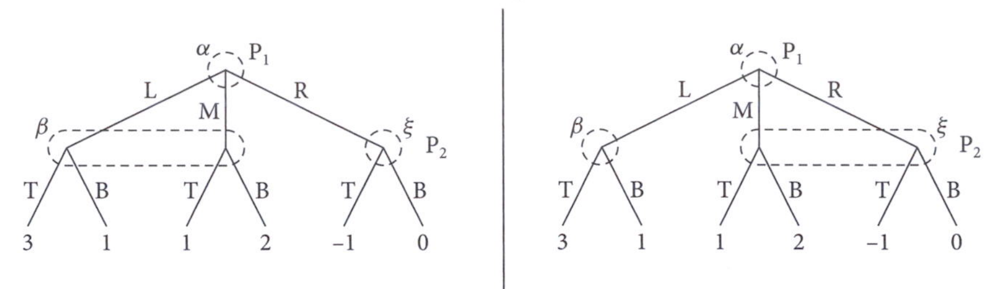

2. Construction of matrix games for the different information sets.

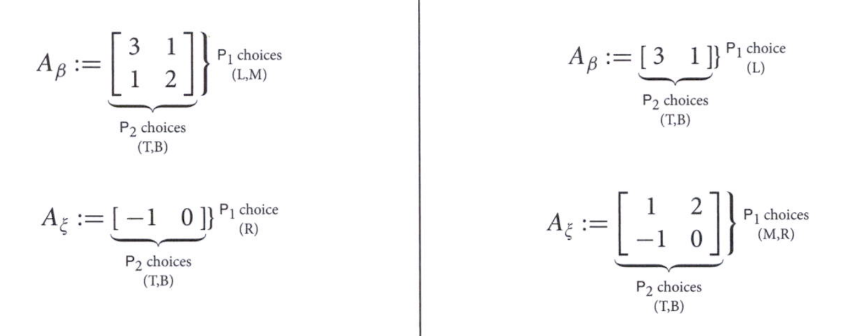

3. Solution of matrix games for the different information sets.

$$\underline{V}(A_\beta) = 1, \quad \bar{V}(A_\beta) = 2,$$

$$\underline{V}(A_\xi) = \bar{V}(A_\xi) = 0$$

The method fails since there is no pure saddle-point equilibrium for A_β.

$$\underline{V}(A_\beta) = \bar{V}(A_\beta) = 3, \quad (i^*, j^*) = (\text{L, T})$$

$$\underline{V}(A_\xi) = \bar{V}(A_\xi) = 0, \quad (i^*, j^*) = (\text{R, B})$$

$$\mathsf{P}_2 \text{ policy} = \begin{cases} \text{T} & \text{if IS} = \beta \\ \text{B} & \text{if IS} = \xi \end{cases}$$

4. Replacement of information sets by the values of the games.

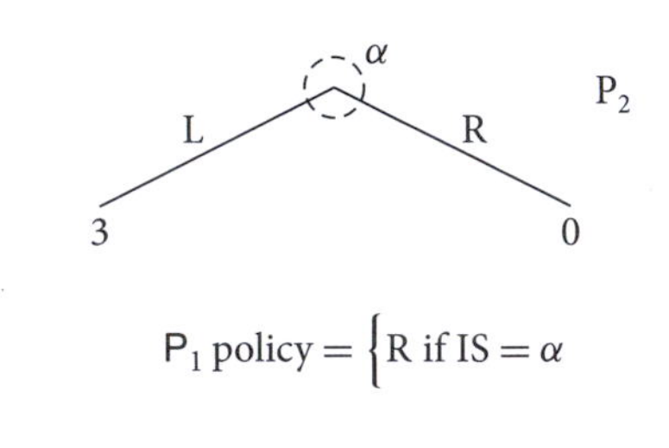

$$\mathsf{P}_1 \text{ policy} = \{\text{R if IS} = \alpha$$

Figure 7.3 Recursive computation of pure saddle-point equilibria for two single-stage games following the procedure described in Section 7.6. As usual, for both games P_1 is the minimizer and P_2 the maximizer.

When all the information sets of a game have a single element, the matrix games corresponding to these sets necessarily have a pure saddle-point equilibrium because one of the players only has a single action. This observation leads to the following consequence of the previous proposition:

Notation. These are called *perfect information games*.

Corollary 7.1. Every zero-sum single-stage game in extensive form for which all information sets have a single element has a pure saddle-point equilibrium. □

Note. See Exercise 7.2. ▷ p. 86

Attention! While the policies constructed with this procedure are guaranteed to be saddle-point equilibria, this procedure may not bring up all possible pure saddle-point equilibria. In fact, in general it will only produce dominating saddle-point equilibria.

Furthermore, when the procedure fails, there may or there may not be pure saddle-point equilibria. This is illustrated in Figure 7.4.

7.7 Feedback Games

Attention! The terminology *feedback games* is somewhat misleading because there are many games in which players can use feedback, but that do not fit this definition.

Feedback games form an important class of multi-stage games, which satisfy two requirements in terms of the information sets allowed.

Definition 7.2 (Feedback games). A multi-stage game in extensive form is said to be a *feedback game (in extensive form)* if the following two conditions hold:

C7.1 no information set spans over multiple stages,

C7.2 the nodes that correspond to the start of each stage are the roots of sub-trees that do not share information sets with each other (including at the level of the root).

Note. C7.1 means that when a player must select an action, they always know the current stage of the game.

Note. C7.2 means that both players have full information about what both players did in past stages of the game.

Figure 7.5 shows examples of games that satisfy and do not satisfy these conditions.

Note. Every single-stage game is a feedback game.

Exercise. Convince yourself that chess is a feedback game.

7.8 Feedback Saddle-Point for Multi-Stage Games

Since in feedback games information sets do not span over multiple stages, we can decompose a policy for a particular player into several sub-policies, one for each stage.

Consider a feedback game with K stages and let us denote by $\mathcal{I}$ the set of all information sets for player P_i. We can partition $\mathcal{I}$ into K disjoint subsets

$$\mathcal{I}_1, \mathcal{I}_2, \dots, \mathcal{I}_K, \qquad \mathcal{I} = \bigcup_{i=1}^{K} \mathcal{I}_i,$$

each one containing all the information sets for a specific stage. If we are then given a policy

$$\gamma: \mathcal{I} \to \mathcal{A}, \qquad \alpha \mapsto \gamma(\alpha)$$

1. Initial games.

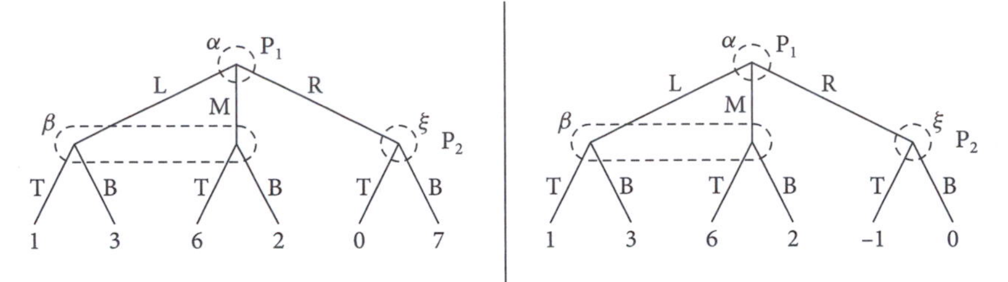

2. Construction of matrix games for the different information sets.

$$A_\beta := \underbrace{\begin{bmatrix} 1 & 3 \\ 6 & 2 \end{bmatrix}}_{\substack{\mathsf{P}_2 \text{ choices} \\ (\mathrm{T,B})}} \Big\} \substack{\mathsf{P}_1 \text{ choices} \\ (\mathrm{L,M})}$$

$$A_\xi := \underbrace{\begin{bmatrix} 0 & 7 \end{bmatrix}}_{\substack{\mathsf{P}_2 \text{ choices} \\ (\mathrm{T,B})}} \} \substack{\mathsf{P}_1 \text{ choice} \\ (\mathrm{R})}$$

$$A_\beta := \underbrace{\begin{bmatrix} 1 & 3 \\ 6 & 2 \end{bmatrix}}_{\substack{\mathsf{P}_2 \text{ choices} \\ (\mathrm{T,B})}} \Big\} \substack{\mathsf{P}_1 \text{choices} \\ (\mathrm{L,M})}$$

$$A_\xi := \underbrace{\begin{bmatrix} -1 & 0 \end{bmatrix}}_{\substack{\mathsf{P}_2 \text{ choices} \\ (\mathrm{T,B})}} \} \substack{\mathsf{P}_1 \text{ choice} \\ (\mathrm{R})}$$

3. Solution of matrix games for the different information sets.

$$\underline{V}(A_\beta) = 2, \qquad \bar{V}(A_\beta) = 3,$$

$$\underline{V}(A_\xi) = \bar{V}(A_\xi) = 7$$

The method fails since there is no pure saddle-point equilibrium for A_β. This game has no pure saddle-point equilibria.

$$\underline{V}(A_\beta) = 2, \qquad \bar{V}(A_\beta) = 3,$$

$$\underline{V}(A_\xi) = \bar{V}(A_\xi) = 0$$

The method fails since there is no pure saddle-point equilibrium for A_β. However, this game has several pure saddle-point equilibria, including the following:

$$\mathsf{P}_2 \text{ policy} = \begin{cases} \mathrm{T} & \text{if IS} = \beta \\ \mathrm{B} & \text{if IS} = \xi \end{cases}$$

$$\mathsf{P}_1 \text{ policy} = \{\, \mathrm{R} \quad \text{if IS} = \alpha$$

This happens essentially because P_1 will never choose L or M since R strictly dominates both choices.

Figure 7.4 Recursive computation of pure saddle-point equilibria for two single-stage games following the procedure described in Section 7.6. As usual, for both games P_1 is the minimizer and P_2 the maximizer.

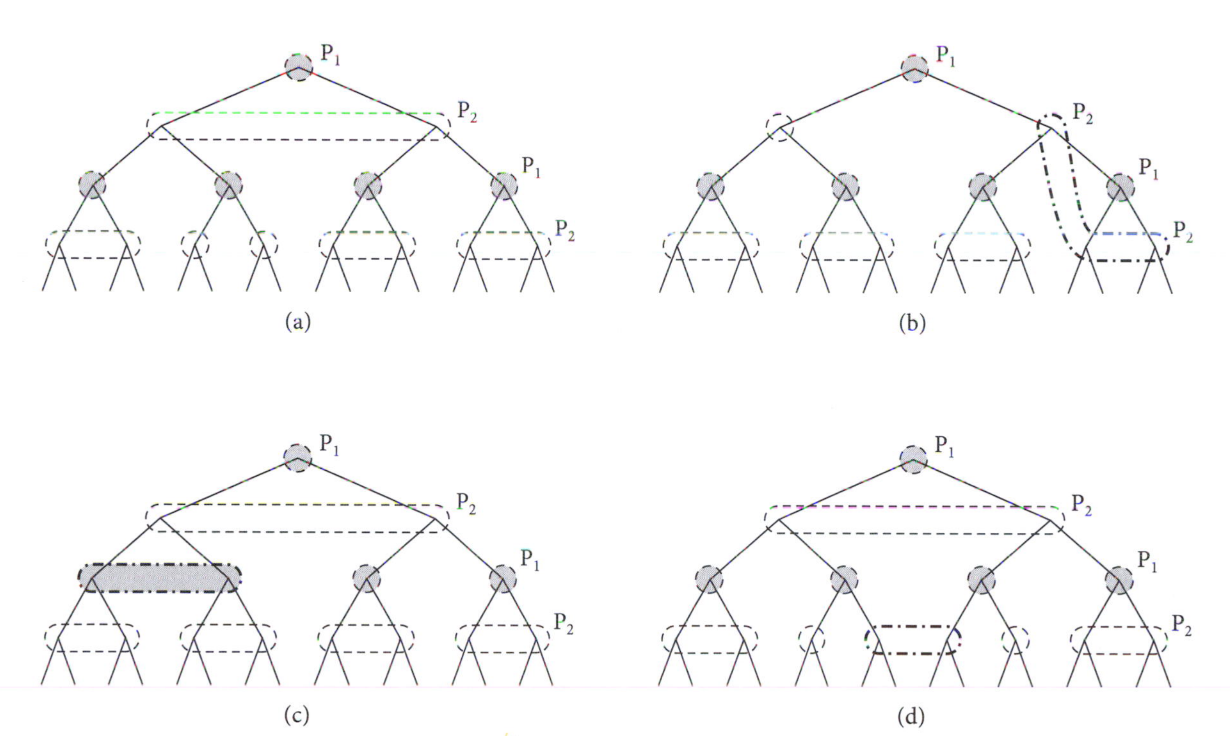

Figure 7.5 Examples of feedback and not feedback multi-stage games. The information sets in gray correspond to the start of a new stage. (a) Feedback game. (b) Not a feedback game because the dash-dot information set spans over multiple stages, violating condition C7.1. (c) Not a feedback game because at the start of the second stage P_1 does not necessarily know the action selected by P_2 in the first stage. The dash-dot information set violates condition C7.2. (d) Not a feedback game because at the second stage, P_2 does not necessarily know the action selected by P_1 at the first stage. The dash-dot information set violates condition C7.2.

that maps each information set in $\mathcal{I}$ to some action in the action space $\mathcal{A}$, we can decompose γ into K sub-policies $\gamma := \{\gamma_1, \gamma_2, \dots, \gamma_K\}$, where each

Notation. In the sequel, we use the shorthand notation $\gamma_{i\dots j}$ to denote the list of sub-policies $\gamma_i, \gamma_{i+1}, \dots, \gamma_j$.

$$\gamma_i : \mathcal{I}_i \to \mathcal{A}, \qquad \alpha \mapsto \gamma(\alpha)$$

only maps the ith stage information sets to actions in $\mathcal{A}$. This decomposition motivates a "new" definition of saddle-point equilibria for feedback games.

Definition 7.3 (Feedback pure saddle-point equilibrium). For a feedback game with K stages, a pair of policies (γ^*, σ^*) with

$$\gamma^* := \{\gamma_1^*, \gamma_2^*, \dots, \gamma_K^*\}, \qquad \sigma^* := \{\sigma_1^*, \sigma_2^*, \dots, \sigma_K^*\},$$

is a *feedback pure saddle-point equilibrium* if for every stage k and every policy

Note. In words, (7.7)–(7.8) mean that no matter what the players do *before stage k* (rational or not) and assuming that both players will play at the equilibrium *after stage* k, the stage k sub-policies (γ_k^*, σ_k^*) must be a pure saddle-point equilibrium.

$\gamma_1, \gamma_2, \ldots, \gamma_{k-1}$ and $\sigma_1, \sigma_2, \ldots, \sigma_{k-1}$ for the stages prior to k we have that

$$J\left(\left\{\gamma_{1\ldots k-1}, \boxed{\gamma_k^*}, \gamma_{k+1\ldots K}^*\right\}, \left\{\sigma_{1\ldots k-1}, \boxed{\sigma_k^*}, \sigma_{k+1\ldots K}^*\right\}\right)$$
$$\leq J\left(\left\{\gamma_{1\ldots k-1}, \boxed{\gamma_k}, \gamma_{k+1\ldots K}^*\right\}, \left\{\sigma_{1\ldots k-1}, \boxed{\sigma_k^*}, \sigma_{k+1\ldots K}^*\right\}\right), \quad \forall \gamma_k \tag{7.7}$$
$$J\left(\left\{\gamma_{1\ldots k-1}, \boxed{\gamma_k^*}, \gamma_{k+1\ldots K}^*\right\}, \left\{\sigma_{1\ldots k-1}, \boxed{\sigma_k^*}, \sigma_{k+1\ldots K}^*\right\}\right)$$
$$\geq J\left(\left\{\gamma_{1\ldots k-1}, \boxed{\gamma_k^*}, \gamma_{k+1\ldots K}^*\right\}, \left\{\sigma_{1\ldots k-1}, \boxed{\sigma_k}, \sigma_{k+1\ldots K}^*\right\}\right), \quad \forall \sigma_k, \tag{7.8}$$

where the universal quantifications refer to all possible k stage pure sub-policies.

Note 8. We shall see shortly that Definition 7.3 is closely related to the basic definition for pure saddle-point equilibria that we have been discussing until now. One may wonder, why introduce a new notion of pure saddle-point equilibria? ▹ p. 83

The use of the terminology "feedback pure saddle-point equilibrium" is justified by the fact that these equilibria are always pure saddle-point equilibrium in the sense of the previous Definition 7.1.

Lemma 7.1. For every feedback game in extensive form, a *feedback pure saddle-point equilibrium* in the sense of Definition 7.3 is always a *pure saddle-point equilibrium* in the sense of Definition 7.1. □

Attention! The converse of this statement is not true, as there may be pure saddle-point equilibria that are not feedback pure saddle-point equilibria.

Proof of Lemma 7.1. To prove this result, we need to show that (7.7)–(7.8) actually imply that

$$J(\gamma^*, \sigma^*) \leq J(\gamma, \sigma^*), \quad \forall \gamma \qquad J(\gamma^*, \sigma^*) \geq J(\gamma^*, \sigma), \quad \forall \sigma, \tag{7.9}$$

which can be re-written in terms of sub-policies as:

$$J\Big(\underbrace{\{\gamma_{1\ldots K}^*\}}_{\gamma^*}, \underbrace{\{\sigma_{1\ldots K}^*\}}_{\sigma^*}\Big) \leq J\Big(\underbrace{\{\gamma_{1\ldots K}\}}_{\gamma}, \underbrace{\{\sigma_{1\ldots K}^*\}}_{\sigma^*}\Big), \quad \forall \gamma_1, \gamma_2, \ldots, \gamma_K$$
$$J\Big(\underbrace{\{\gamma_{1\ldots K}^*\}}_{\gamma^*}, \underbrace{\{\sigma_{1\ldots K}^*\}}_{\sigma^*}\Big) \leq J\Big(\underbrace{\{\gamma_{1\ldots K}^*\}}_{\gamma^*}, \underbrace{\{\sigma_{1\ldots K}\}}_{\sigma}\Big), \quad \forall \sigma_1, \sigma_2, \ldots, \sigma_K.$$

To accomplish this, we start by using (7.7) for the first stage $k = 1$:

$$J\Big(\left\{\boxed{\gamma_1^*}, \gamma_{2\ldots K}^*\right\}, \underbrace{\left\{\boxed{\sigma_1^*}, \sigma_{2\ldots K}^*\right\}}_{\sigma^*}\Big) \leq J\Big(\left\{\boxed{\gamma_1}, \gamma_{2\ldots K}^*\right\}, \underbrace{\left\{\boxed{\sigma_1^*}, \sigma_{2\ldots K}^*\right\}}_{\sigma^*}\Big), \quad \forall \gamma_1.$$

We then use (7.7) for the next stage $k = 2$ and with $\sigma_1 := \sigma_1^*$ in the first stage

$$J\Bigg(\Big\{\gamma_1, \boxed{\gamma_2^*}, \gamma_{3\ldots K}^*\Big\}, \underbrace{\Big\{\sigma_1^*, \boxed{\sigma_2^*}, \sigma_{3\ldots K}^*\Big\}}_{\sigma^*}\Bigg)$$

$$\leq J\Bigg(\Big\{\gamma_1, \boxed{\gamma_2}, \gamma_{3\ldots K}^*\Big\}, \underbrace{\Big\{\sigma_1^*, \boxed{\sigma_2^*}, \sigma_{3\ldots K}^*\Big\}}_{\sigma^*}\Bigg), \quad \forall \gamma_2.$$

Combining these two inequalities we obtain

$$J\Bigg(\Big\{\gamma_{1\ldots K}^*\Big\}, \underbrace{\Big\{\sigma_{1\ldots K}^*\Big\}}_{\sigma^*}\Bigg) \leq J\Bigg(\Big\{\gamma_1, \gamma_2, \gamma_{3\ldots K}^*\Big\}, \underbrace{\Big\{\sigma_{1\ldots K}^*\Big\}}_{\sigma^*}\Bigg), \quad \forall \gamma_1, \gamma_2.$$

As we combine this with (7.7) for each subsequent stage $k \in \{3, 4, \ldots, K\}$ and with $\sigma_i = \sigma_i^*, \forall i \in \{1, 2, \ldots, k-1\}$ for the previous stages, we eventually obtain

$$J\Bigg(\Big\{\gamma_{1\ldots K}^*\Big\}, \underbrace{\Big\{\sigma_{1\ldots K}^*\Big\}}_{\sigma^*}\Bigg) \leq J\Bigg(\Big\{\gamma_{1\ldots K}\Big\}, \underbrace{\Big\{\sigma_{1\ldots K}^*\Big\}}_{\sigma^*}\Bigg), \quad \forall \gamma_1, \gamma_2, \ldots, \gamma_K,$$

which proves that $J(\gamma^*, \sigma^*) \leq J(\gamma, \sigma^*)$, $\forall \gamma$. The other saddle-point inequality in (7.9) can be proved similarly using (7.8). ∎

Note 8 (Feedback pure saddle-point equilibria). One may wonder, why introduce a new notion of saddle-point equilibria when we already had a very reasonable one. There are several reasons for this (listed in "decreasing order of importance"):

1. Finding feedback saddle-point equilibria is actually easier than finding saddle-point equilibria that are not feedback. We shall see this shortly in Section 7.9.
2. Feedback saddle-point equilibria are "better" in the sense that they still provide optimal security levels, even if the player did not play rationally in the past. Of course, these security levels may not be as good as the ones that could have been obtained if the player had been playing rationally since the beginning of the game.

Note. Recall that for zero-sum games played by rational players all saddle-point equilibria have the same value and are completely interchangeable.

When we restrict our attention to pure policies, there may be feedback games that have saddle-point equilibria, but that do not have feedback pure saddle-point equilibria (cf. example in Figure 7.4.). Therefore we may end up without equilibria by restricting our search to feedback pure saddle-point equilibria.

However, we will shortly see that, when we allow for stochastic policies, all these games will have feedback saddle-point equilibria so the feedback requirement will not reduce the class of games for which we can find equilibria. □

7.9 Recursive Computation of Equilibria for Multi-Stage Games

Notation. We shall later find essentially the same procedure under the name of *dynamic programming* (cf. Lecture 17). ▷ p. 201

The procedure that we discussed in Section 7.6 to compute pure saddle-point equilibria for single-stage games can be generalized to compute *feedback* pure saddle-point equilibria for *multi-stage games*. For simplicity, in listing the steps of this procedure we assume that P_1 is the "first-acting" player at every stage and that the game has exactly K stages.

1. Construct a matrix game corresponding to each information set of P_2 at the Kth stage. Each matrix game will have one action of P_1 for each edge entering the information set and one action of P_2 for each edge leaving the information set.

Note. If there are multiple pure saddle-points, any one will do.

2. Compute pure saddle-point equilibria for each of the matrix games constructed.

   If any of these matrix games does not have a pure saddle-point equilibrium, this method fails. Otherwise, P_2's pure security/saddle-point sub-policy σ_K^* for the Kth stage should map to each information set the action corresponding to P_2's security policy/saddle-point in the corresponding matrix game.

Note. Recall that the sub-policy for the Kth stage is a function that maps each information set at that stage with one action.

Note. See Figure 7.6. ▹ p. 85

3. Replace each information set for P_2 at the Kth stage by the value of the corresponding pure saddle-point equilibrium. The link leading to this value should be labeled with P_1's pure security/saddle-point in the corresponding matrix game.

Note. If multiple policies achieve the same value, any one will do.

4. The pure security/saddle-point sub-policy γ_K^* for the player P_1 at the Kth stage maps to each information set of P_1, the action in the link corresponding to the most favorable pure saddle-point value for P_1.

   The two policies (γ_K^*, σ_K^*) so computed have the property that, for every policy $\gamma_1, \gamma_2, \ldots, \gamma_{K-1}$ and $\sigma_1, \sigma_2, \ldots, \sigma_{K-1}$ for the stages prior to K,

$$J\left(\left\{\gamma_{1\ldots K-1}, \boxed{\gamma_K^*}\right\}, \left\{\sigma_{1\ldots K-1}, \boxed{\sigma_K^*}\right\}\right)$$
$$\leq J\left(\left\{\gamma_{1\ldots K-1}, \boxed{\gamma_K}\right\}, \left\{\sigma_{1\ldots K-1}, \boxed{\sigma_K^*}\right\}\right), \quad \forall \gamma_K$$
$$J\left(\left\{\gamma_{1\ldots K-1}, \boxed{\gamma_K^*}\right\}, \left\{\sigma_{1\ldots K-1}, \boxed{\sigma_K^*}\right\}\right)$$
$$\geq J\left(\left\{\gamma_{1\ldots K-1}, \boxed{\gamma_K^*}\right\}, \left\{\sigma_{1\ldots K-1}, \boxed{\sigma_K}\right\}\right), \quad \forall \sigma_K,$$

   as required for a feedback pure saddle-point equilibrium.

Note. See Figure 7.6. ▹ p. 85

5. Replace each information set for P_1 at the Kth stage by the value corresponding to the link selected by P_1.

   At this point we have a game with $K-1$ stages, and we compute the sub-policies $(\gamma_{K-1}^*, \sigma_{K-1}^*)$ using the exact same procedure described above. This algorithm is iterated until all sub-policies have been obtained. Moreover, at every stage, we guarantee by construction that the conditions (7.7)–(7.8) for a feedback pure saddle-point equilibrium hold.

Attention! To reduce the number of stages in the game, we replace lower-level stages by the values obtained using the computed equilibrium sub-policies, which is precisely consistent with (7.7)–(7.8).

Proposition 7.3. For any zero-sum game in extensive form with a finite number of stages, if every matrix game constructed using the above procedure has a pure saddle-point equilibrium, then the overall game in extensive form has a feedback pure saddle-point equilibrium. □

When all the information sets of a game have a single element, the matrix games corresponding to these sets necessarily have a saddle-point equilibrium because one of the players only has a single action. This observation leads to the following important consequence of the previous proposition:

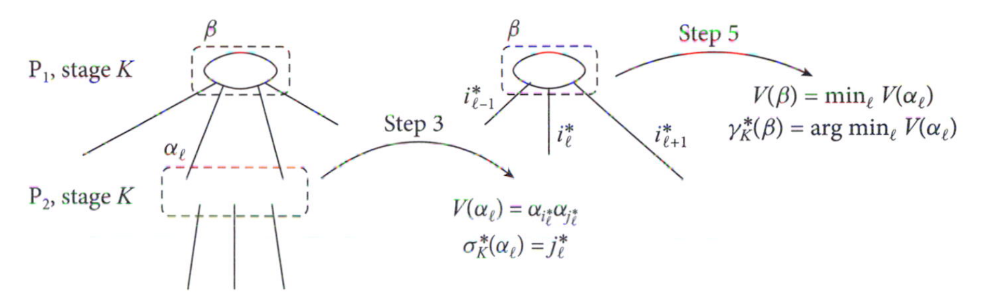

Figure 7.6 Illustration of steps 3–5 in the recursive computation of equilibria for multi-stage games. In this diagram, α denotes the information state corresponding to player P_2 at the Kth stage, $V(\alpha)$ the value of the matrix game associated with that information state, and (i^*, j^*) the corresponding saddle-point equilibrium point.

Notation. These are called *perfect information games.*

Corollary 7.2. Every zero-sum game in extensive form with a finite number of stages for which all information sets have a single element has a feedback pure saddle-point equilibrium. □

7.10 Practice Exercise

7.1 (Number of games in extensive form). How many different extensive games are possible for a given $m \times n$ matrix of outcomes?

Solution to Exercise 7.1. Assuming that player P_1 plays first, the number of distinct games is given by the *Bell number* b_m, which is equal to the number of partitions of the set $\{1, 2, \ldots, m\}$ and can be computed recursively by

$$b_{k+1} = \sum_{i=1}^{k} \binom{k}{i}$$

(cf. Table 7.1). On the other hand, assuming that player P_2 plays first, the number of distinct games is given by the *Bell number* b_n. Consequently, the total number of distinct games is equal to

$$b_m - b_n - 1,$$

where the subtraction by 1 is to avoid double counting the simultaneous play, which otherwise would be counted twice. For example for the 2×3 matrix in (7.1) we have $b_2 + b_3 - 1 = 2 + 5 - 1 = 6$ games.

TABLE 7.1 Bell Numbers

m	0	1	2	3	4	5	6	7	8	9	...	15
b_m	1	1	2	5	15	52	203	877	4149	21147	...	$\approx 1.4 \times 10^9$

7.11 Additional Exercises

7.2. Show that the procedure described in Section 7.6 may not find all possible pure saddle-point equilibria by considering the $P_1 - P_2$ game in extensive form in Figure 7.1 (see also Example 7.1).

7.3 (Chess). Justify that the game of chess has a feedback pure saddle-point equilibrium.

Hint: You are not being asked to find a pure saddle-point equilibrium, just to show that it must exist.

LECTURE 8

Stochastic Policies for Games in Extensive Form

This lecture introduces two types of stochastic policies for games in extensive form and discusses the existence and computation of saddle-point equilibria.

8.1 Mixed Policies and Saddle-Point Equilibria

Take a game in extensive form for which

$$\Gamma_1 = \{\gamma_1, \gamma_2, \dots, \gamma_m\}, \qquad \Gamma_2 = \{\sigma_1, \sigma_2, \dots, \sigma_n\},$$

are the finite sets of all pure policies for players P_1 and P_2, respectively. As we saw in Section 7.5, this game can be represented in matrix form by an $m \times n$ matrix A_{ext}, whose entry a_{ij} is equal to the outcome of the game $J(\gamma_i, \sigma_j)$ when P_1 selects the pure policy γ_i corresponding to the ith row of A_{ext} and P_2 selects the pure policy σ_j corresponding to the jth column.

For games in extensive form, a *mixed policy* corresponds to selecting a *pure policy* randomly according to a previously selected probability distribution before the game starts, and then playing that policy throughout the game:

- A *mixed policy for* P_1 is a set of numbers

$$\{y_1, y_2, \dots, y_m\}, \quad \sum_{i=1}^{m} y_i = 1, \quad y_i \geq 0, \quad \forall i \in \{1, 2, \dots, m\},$$

with the understanding that y_i is the probability that P_1 uses to select the pure policy γ_i.

- A *mixed policy for* P_2 is a set of numbers

$$\{z_1, z_2, \ldots, z_n\}, \quad \sum_{j=1}^{n} z_j = 1, \quad z_j \geq 0, \quad \forall j \in \{1, 2, \ldots, n\},$$

with the understanding that z_j is the probability that P_2 uses to select the pure policy σ_j.

It is assumed that the random selections by both players are done statistically independently and the players will try to optimize the *expected outcome of the game*, which is given by

Note. See Section 4.1 on how to derive this expression.

$$J = \sum_{i,j} J(\gamma_i, \sigma_j) \text{ Prob}(\mathsf{P}_1 \text{ selects } \gamma_i \text{ and } \mathsf{P}_2 \text{ selects } \sigma_j) = y' A_{\text{ext}} z,$$

where $y := [y_1 \; y_2 \; \cdots \; y_m]'$ and $z := [z_1 \; z_2 \; \cdots \; z_n]'$.

Having defined mixed policies for a game in extensive form, we can use the concepts of security levels, security policies, and saddle-point equilibria introduced in Section 4.5 for general games, with the understanding that:

1. the *action spaces* are the sets $\mathcal{Y}$ and $\mathcal{Z}$ of all mixed policies for players P_1 and P_2, respectively; and
2. for a particular pair of mixed policies $y \in \mathcal{Y}, z \in \mathcal{Z}$ the *outcome of the game* when player P_1 uses policy y and P_2 uses policy z is given by $J(y, z) := y' A_{\text{ext}} z$.

In particular, this leads to the following definition of saddle-point equilibria:

Attention! With the definitions in Section 4.5 we recover all the results in Proposition 4.2 regarding the relationship between security levels/policies and saddle-point equilibria. However, here the supremum and infimum are minimum and maximum, respectively, because we are optimizing continuous functions over compact sets.

Definition 8.1 (Mixed saddle-point equilibrium). A pair of policies $(y^*, z^*) \in \mathcal{Y} \times \mathcal{Z}$ is called a *mixed saddle-point equilibrium* if

$$y^{*\prime} A_{\text{ext}} z^* \leq y' A_{\text{ext}} z^*, \quad \forall y \in \mathcal{Y}, \qquad y^{*\prime} A_{\text{ext}} z^* \geq y^{*\prime} A_{\text{ext}} z, \quad \forall z \in \mathcal{Z}$$

and $y^{*\prime} A_{\text{ext}} z^*$ is called the *saddle-point value*.

The following is a straightforward consequence of Proposition 4.2 and the Minimax Theorem 5.1:

Corollary 8.1. For every zero-sum game in extensive form with (finite) matrix representation A_{ext}:

Notation. Since the two security levels are always equal, one just writes $V_m(A_{\text{ext}})$, which is simply called the *(mixed) value of the game.*

P8.1 A mixed saddle-point equilibrium always exists and

$$\underline{V}_m(A_{\text{ext}}) := \max_{z \in \mathcal{Z}} \min_{y \in \mathcal{Y}} y' A_{\text{ext}} z = \min_{y \in \mathcal{Y}} \max_{z \in \mathcal{Z}} y' A_{\text{ext}} z =: \bar{V}_m(A_{\text{ext}}). \quad (8.1)$$

P8.2 If y^* and z^* are mixed security policies for P_1 and P_2, respectively then (y^*, z^*) is a mixed saddle-point equilibrium and its value $y^{*\prime} A_{\text{ext}} z^*$ is equal to (8.1).

P8.3 If (y^*, z^*) is a mixed saddle-point equilibrium then y^* and z^* are mixed security policies for P_1 and P_2, respectively and (8.1) is equal to the mixed saddle-point value $y^{*\prime} A_{\text{ext}} z^*$.

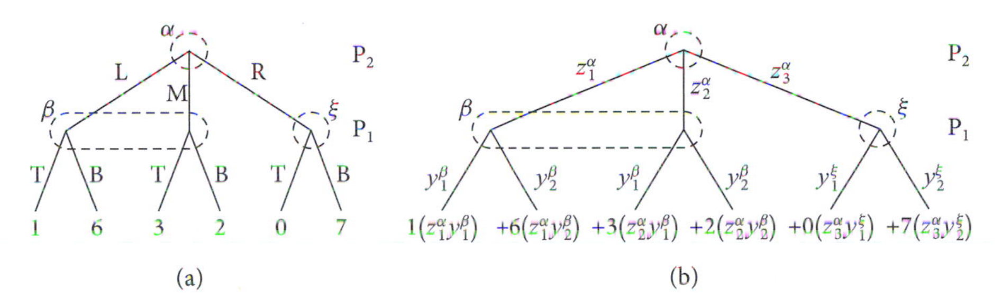

Figure 8.1 Game in extensive form for Examples 8.1 and 8.2. (a) Extensive form; (b) Computation of expected outcome for behavioral policies.

P8.4 If (y_1^*, z_1^*) and (y_2^*, z_2^*) are mixed saddle-point equilibria then (y_1^*, z_2^*) and (y_2^*, z_1^*) are also mixed saddle-point equilibria and

$$y_1^{*\prime} A_{\text{ext}} z_1^* = y_2^{*\prime} A_{\text{ext}} z_2^* = y_1^{*\prime} A_{\text{ext}} z_2^* = y_2^{*\prime} A_{\text{ext}} z_1^*. \qquad \square$$

Example 8.1. Consider the game in extensive form in Figure 8.1 and suppose that we enumerate the policies available for P_1 and P_2 as follows:

P_1: *IS*	γ_1	γ_2	γ_3	γ_4
β	T	T	B	B
ξ	T	B	T	B

P_2: *IS*	σ_1	σ_2	σ_3
α	L	M	R

Note. See Exercise 8.1. ▷ p. 96

In this case, we obtain

$$A_{\text{ext}} = \underbrace{\begin{bmatrix} 1 & 3 & 0 \\ 1 & 3 & 7 \\ 6 & 2 & 0 \\ 6 & 2 & 7 \end{bmatrix}}_{\substack{P_2 \text{ policies} \\ (\sigma_1, \sigma_2, \sigma_3)}} \Bigg\} \substack{P_1 \text{ policies} \\ (\gamma_1, \gamma_2, \gamma_3, \gamma_4)}$$

Exercise. Are these the only security policies? See Exercise 8.2. ▷ p. 97

Note. Why would P_1 ever use policy γ_4 in view of the fact that this policy is never better than γ_3 and is sometimes strictly worse? Because a rational P_2 will never choose R and therefore γ_3 and γ_4 lead to the same outcome. However, γ_3 is a dominating policy so one would generally prefer this policy over γ_4, in spite of the fact that γ_3 is more "complex" since it involves a conditional decision.

with the following value and security policies:

$$V_m(A_{\text{ext}}) = \frac{8}{3}, \qquad y^* = \begin{bmatrix} \frac{2}{3} \\ 0 \\ 0 \\ \frac{1}{3} \end{bmatrix} \text{ or } \begin{bmatrix} \frac{2}{3} \\ 0 \\ \frac{1}{3} \\ 0 \end{bmatrix}, \qquad z^* = \begin{bmatrix} \frac{1}{6} \\ \frac{5}{6} \\ 0 \end{bmatrix}. \tag{8.2}$$

This corresponds to the following choices for the players before the game starts:

P_1 selects pure policy $\begin{cases} \gamma_1 & \text{with probability } \frac{2}{3} \text{ (T)} \\ \gamma_4 & \text{with probability } \frac{1}{3} \text{ (B)} \end{cases}$ or $\begin{cases} \gamma_1 & \text{with probability } \frac{2}{3} \text{ (T)} \\ \gamma_3 & \text{with probability } \frac{1}{3} \text{ (B or T)} \end{cases}$

P_2 selects pure policy $\begin{cases} \sigma_1 & \text{with probability } \frac{1}{6} \text{ (L)} \\ \sigma_2 & \text{with probability } \frac{5}{6} \text{ (M)} \end{cases}$

□

8.2 Behavioral Policies for Games in Extensive Form

Behavioral policies also involve randomization but the key difference with respect to mixed policies is that the randomization is done over actions as the game is played and not over pure policies before the game starts.

A *behavioral policy* for the player P_i is a decision rule π_i (i.e., a function) that associates to each information set α of P_i a probability distribution $\pi_i(\alpha)$ over the possible actions for that information set. When P_i is in a particular information set α the distribution $\pi_i(\alpha)$ is used to decide on an action for P_i.

Note. With behavioral policies, the randomization over actions is made as the game progresses, as a player enters a particular information set.

It is assumed that the random selections by both players at the different information sets are all done statistically independently and that the players try to optimize the resulting *expected outcome* J of the game.

As for pure policies, we can perform a per-stage decomposition of behavioral policies for feedback games. In particular, if we are given a behavioral policy γ that maps information sets to probability distributions over actions, we can decompose γ into K sub-policies

$$\gamma := \{\gamma_1, \gamma_2, \ldots, \gamma_K\},$$

where each γ_i only maps the ith stage information sets to probability distributions over actions.

Example 8.2. Consider again the game in extensive form in Figure 8.1. A behavioral policy for P_2 is a function that maps the (only) information set α into a probability distribution as follows:

P_2:

IS	L	M	R
α	z_1^α	z_2^α	z_3^α

$$z_1^\alpha, z_2^\alpha, z_3^\alpha \geq 0, z_1^\alpha, z_2^\alpha, z_3^\alpha = 1,$$

and a behavioral policy for P_1 is a function that maps each of the information sets β, ξ into a probability distribution as follows:

Attention! The player may use different probability distributions to select actions in different information sets.

P_1:

IS	T	B
β	y_1^β	y_2^β
ξ	y_1^ξ	y_2^ξ

$$y_1^\beta, y_2^\beta \geq 0, y_1^\beta, z_2^\beta = 1$$
$$y_1^\xi, y_2^\xi \geq 0, y_1^\xi, z_2^\xi = 1.$$

For these behavioral policies, the expected outcome of the game is given by the following expression

$$J = 1\, z_1^\alpha y_1^\beta + 6\, z_1^\alpha y_2^\beta + 3\, z_2^\alpha y_1^\beta + 2\, z_2^\alpha y_2^\beta + 0\, z_3^\alpha y_1^\beta + 7\, z_3^\alpha y_2^\beta$$

(see Figure 8.1(b)). □

Note 9. There is a simple systematic procedure to compute the expected outcome J for arbitrary games in extensive form under behavioral policies. ▷ p. 90

Note 9. The following systematic procedure can be used to compute the expected outcome for an arbitrary game in extensive form under behavioral policies:

1. Label every link of the tree describing the game in extensive form with the probability with which that action will be chosen by the behavioral policy of the corresponding player.

2. For each leaf multiply all the probabilities of the links that connect the root to that leaf. The resulting number is the probability for that outcome.
3. The expected reward is obtained by summing over all leaves the product of the leaf's outcome by its probability computed in 2. □

8.3 Behavioral Saddle-Point Equilibria

Having defined behavioral policies for a game in extensive form, we can use the concepts of security levels, security policies, and saddle-point equilibria introduced in Section 4.5 for general games, with the understanding that

1. the *action spaces* are the sets $\Gamma_1^{\flat}$ and $\Gamma_2^{\flat}$ of all behavioral policies for players P_1 and P_2, respectively; and
2. for a particular pair of behavioral policies $\gamma \in \Gamma_1^{\flat}, \sigma \in \Gamma_2^{\flat}$, we denote by $J(y, z)$ the *expected outcome of the game* when player P_1 uses policy γ and P_2 uses policy σ.

Attention! With the definitions in Section 4.5 we recover all the results in Proposition 4.2 regarding the relationship between security levels/policies and saddle-point equilibria.

This leads to the following definition of saddle-point equilibrium:

Definition 8.2 (Behavioral saddle-point equilibrium). A pair of policies $(\gamma^*, \sigma^*) \in \Gamma_1^{\flat} \times \Gamma_2^{\flat}$ is a *behavioral saddle-point equilibrium* if

$$J(\gamma^*, \sigma^*) \leq J(\gamma, \sigma^*), \quad \forall \gamma \in \Gamma_1^{\flat}, \qquad J(\gamma^*, \sigma^*) \geq J(\gamma^*, \sigma), \quad \forall \sigma \in \Gamma_2^{\flat}$$

and $J(\gamma^*, \sigma^*)$ is called the behavioral saddle-point value.

For feedback games, we can also extend to behavioral policies the notion of feedback saddle-point equilibrium that we used before for pure policies:

Notation. We use the shorthand notation $\gamma_{i \ldots j}$ to denote the list of sub-policies $\gamma_i, \gamma_{i+1}, \ldots, \gamma_j$.

Definition 8.3 (Feedback behavioral saddle-point equlibrium). For a feedback game with K stages, a pair of policies (γ^*, σ^*) with

$$\gamma^* := \{\gamma_1^*, \gamma_2^*, \ldots, \gamma_K^*\}, \qquad \sigma^* := \{\sigma_1^*, \sigma_2^*, \ldots, \sigma_K^*\},$$

is a *feedback behavioral saddle-point equilibrium* if for every stage k and all policies $\gamma_1, \gamma_2, \ldots, \gamma_{k-1}$ and $\sigma_1, \sigma_2, \ldots, \sigma_{k-1}$ for the stages prior to k we have that

Note. In words, this means that no matter what the players do *before stage* k (rational or not) and assuming that both players will play at the equilibrium *after stage* k, the stage k sub-policies (γ_k^*, σ_k^*) must be a saddle-point equilibrium.

$$\begin{aligned} J\Big(\{\gamma_{1\ldots k-1}, \boxed{\gamma_k^*}, \gamma_{k+1\ldots K}^*\}, \{\sigma_{1\ldots k-1}, \boxed{\sigma_k^*}, \sigma_{k+1\ldots K}^*\}\Big) \\ \leq J\Big(\{\gamma_{1\ldots k-1}, \boxed{\gamma_k}, \gamma_{k+1\ldots K}^*\}, \{\sigma_{1\ldots k-1}, \boxed{\sigma_k^*}, \sigma_{k+1\ldots K}^*\}\Big), \quad \forall \gamma_k \end{aligned} \tag{8.3}$$

$$\begin{aligned} J\Big(\{\gamma_{1\ldots k-1}, \boxed{\gamma_k^*}, \gamma_{k+1\ldots K}^*\}, \{\sigma_{1\ldots k-1}, \boxed{\sigma_k^*}, \sigma_{k+1\ldots K}^*\}\Big) \\ \geq J\Big(\{\gamma_{1\ldots k-1}, \boxed{\gamma_k^*}, \gamma_{k+1\ldots K}^*\}, \{\sigma_{1\ldots k-1}, \boxed{\sigma_k}, \sigma_{k+1\ldots K}^*\}\Big), \quad \forall \sigma_k, \end{aligned} \tag{8.4}$$

where the universal quantifications refer to all possible k stage behavioral sub-policies.

Note. See Exercise 8.5 regarding the proof of Lemma 8.1. ▷ p. 102

As for pure policies, feedback behavioral saddle-point equilibria are always behavioral saddle-point equilibria.

Lemma 8.1. For every feedback game in extensive form, a *feedback behavioral saddle-point equilibrium* in the sense of Definition 8.3 is always a *behavioral saddle-point equilibrium* in the sense of Definition 8.2. □

Note. This includes, as a special case, every single-stage game in extensive form.

It turns out that every feedback game in extensive form has a feedback behavioral saddle-point equilibrium.

Theorem 8.1. For every zero-sum feedback game G in extensive form with a finite number of stages:

Notation. Since the two security levels are always equal, one just writes $V_\flat(G)$, which is simply called the *(behavioral) value of the game.*

P8.5 A feedback behavioral saddle-point equilibrium always exists and

$$\underline{V}_\flat(G) := \max_{\sigma\in\Gamma_2^\flat} \min_{\gamma\in\Gamma_1^\flat} J(\gamma,\sigma) = \min_{\gamma\in\Gamma_1^\flat} \max_{\sigma\in\Gamma_2^\flat} J(\gamma,\sigma) =: \bar{V}_\flat(G). \qquad (8.5)$$

P8.6 If γ^* and σ^* are behavioral security policies for P_1 and P_2, respectively then (γ^*, σ^*) is a behavioral saddle-point equilibrium and its value $J(\gamma^*, \sigma^*)$ is equal to the (8.5).

P8.7 If (γ^*, σ^*) is a behavioral saddle-point equilibrium, then γ^* and σ^* are behavioral security policies for P_1 and P_2, respectively and (8.5) is equal to the behavioral saddle-point value $J(\gamma^*, \sigma^*)$.

P8.8 If (γ_1^*, σ_1^*) and (γ_2^*, σ_2^*) are behavioral saddle-point equilibria then (γ_1^*, σ_2^*) and (γ_2^*, σ_1^*) are also behavioral saddle-point equilibria and

$$J(\gamma_1^*, \sigma_1^*) = J(\gamma_2^*, \sigma_2^*) = J(\gamma_1^*, \sigma_2^*) = J(\gamma_2^*, \sigma_1^*).$$ □

To prove this Theorem, we will constructively show that a behavioral saddle-point equilibrium always exists, from which P8.5–P8.8 follow using the general results in Proposition 4.2.

8.4 Behavioral vs. Mixed Policies

As we saw in Section 7.4, the total number of distinct pure policies for a given player P_i is given by

$$\underbrace{(\text{\# actions of 1st IS}) \times (\text{\# actions of 2nd IS}) \times \cdots \times (\text{\# actions of last IS})}_{\text{product over all information sets for player } \mathsf{P}_i}.$$

Note. Since mixed policies are probability distributions, once we have decided on the probabilities for all but the last pure policy, the probability of the last one is automatically determined because all probabilities must add up to one.

and therefore mixed policies are probability distributions over these many pure actions, which means that we have

$$\underbrace{(\text{\# actions of 1st IS}) \times (\text{\# actions of 2nd IS}) \times \cdots \times (\text{\# actions of last IS})}_{\text{product over all information sets for player } \mathsf{P}_i} - 1 \qquad (8.6)$$

degrees of *freedom in selecting a mixed policy.*

As for behavioral policies, for each information set we need to select a probability distribution over the actions of that information set. Such a probability distribution has as many degrees of freedom as the number of actions minus one. Therefore, the total number of *degrees of freedom available for the selection of behavioral policies* is given by

$$\underbrace{(\#\text{ actions of 1st IS} - 1) + (\#\text{ actions of 2nd IS} - 1) + \cdots + (\#\text{ actions of last IS} - 1)}_{\text{sum over all information sets for player } \mathsf{P}_i},$$

which is generally a number far smaller than (8.6).

By simply counting degrees of freedom we have seen that the set of mixed strategies is far richer than the set of behavioral strategies. In fact, this set is sufficiently rich so that every game in extensive form has saddle-point equilibria in mixed policies. It turns out that for large classes of games (including the zero-sum feedback games in extensive form considered in Section 8.3), the set of behavioral policies is already sufficiently rich so that one can already find saddle-point equilibria in behavioral policies. Moreover, since the number of degrees of freedom for behavioral policies is much lower, finding such equilibria is computationally much simpler, as we shall see shortly.

8.5 Recursive Computation of Equilibria for Feedback Games

Note. This shows that behavioral saddle-point equilibrium always exist for feedback games, as stated in Proposition 4.2.

Example. Figure 8.2 contains examples illustrating how to apply this procedure. ▹ p. 94

The procedure that we discussed in Section 7.9 to compute *pure* saddle-point equilibria for *feedback multi-stage games* can also be used to compute *behavioral* saddle-point equilibria. The key difference is that now that procedure will never fail and will always find an equilibrium. For simplicity, in listing the steps of this procedure we assume that P_1 is the "first-acting" player at every stage and that the game has exactly K stages.

Note. If there are multiple mixed saddle-points any one will do.

Note. Such saddle-point equilibria always exist because of the Minimax Theorem 5.1.

Note. Recall that the sub-policy for the Kth stage is a function that maps each information set at that stage with one probability distribution over actions.

Note. If multiple policies achieve the same value, any one will do.

1. Construct a matrix game corresponding to each information set of P_2 at the Kth stage. Each matrix game will have one action of P_1 for each edge entering the information set and one action of P_2 for each edge leaving the information set.
2. Compute *mixed* saddle-point equilibria for each of the matrix games constructed.

   P_2's behavioral security/saddle-point sub-policy σ_K^* for the Kth stage should map to each information set the probability distribution over actions corresponding to P_2's mixed security policy in the corresponding matrix game.
3. Replace each information set for P_2 at the Kth stage by the value of the corresponding mixed saddle-point equilibrium. The link leading to this value should be the labeled with P_1's mixed policy for the corresponding matrix game.
4. The behavioral security/saddle-point sub-policy γ_K^* for the player P_1 at the Kth stage maps to each information set of P_1 the probability distribution in the link corresponding to the most favorable saddle-point value for P_1.

   The two policies (γ_K^*, σ_K^*) so computed have the property that, for all policies $\gamma_1, \gamma_2, \ldots, \gamma_{K-1}$ and $\sigma_1, \sigma_2, \ldots, \sigma_{K-1}$ for the stages prior to K,

1. Initial games.

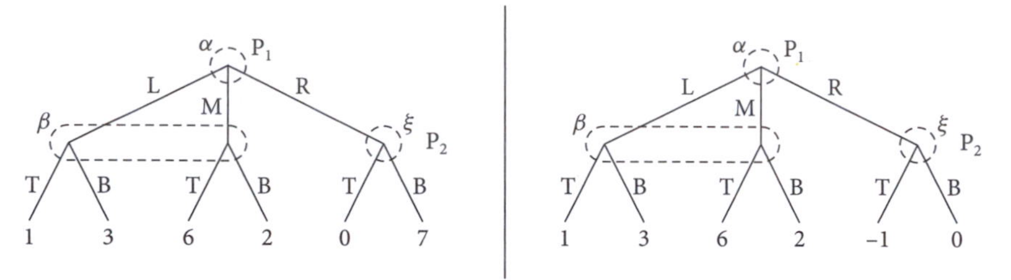

2. Construction of matrix games for the different information sets.

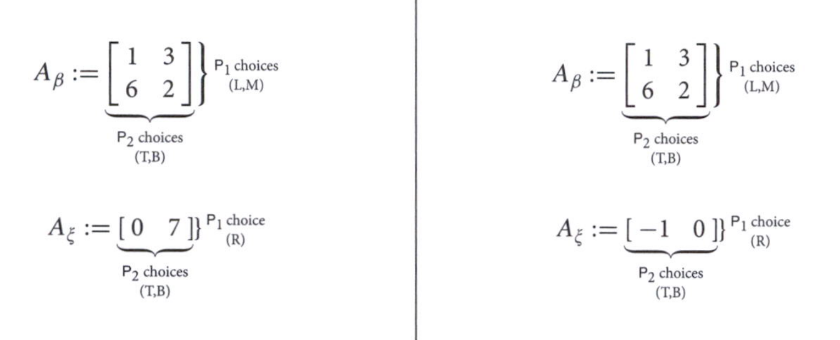

3. Solution of matrix games for the different information sets.

$$V_m(A_\beta) = \frac{8}{3}, \quad y^*_\beta := \begin{bmatrix} \frac{1}{6} \\ \frac{5}{6} \end{bmatrix}, \quad z^*_\beta := \begin{bmatrix} \frac{2}{3} \\ \frac{1}{3} \end{bmatrix}$$

$$V_m(A_\xi) = 7, \quad y^*_\xi := 1, \quad z^*_\xi := \begin{bmatrix} 0 \\ 1 \end{bmatrix}$$

$$\mathsf{P}_2 \text{ policy} = \begin{cases} \begin{bmatrix} \frac{2}{3} \\ \frac{1}{3} \end{bmatrix} & \text{if IS} = \beta \\ \begin{bmatrix} 0 \\ 1 \end{bmatrix} & \text{if IS} = \xi \end{cases}$$

$$V_m(A_\beta) = \frac{8}{3}, \quad y^*_\beta := \begin{bmatrix} \frac{1}{6} \\ \frac{5}{6} \end{bmatrix}, \quad z^*_\beta := \begin{bmatrix} \frac{2}{3} \\ \frac{1}{3} \end{bmatrix}$$

$$V_m(A_\xi) = 0, \quad y^*_\xi := 1, \quad z^*_\xi := \begin{bmatrix} 0 \\ 1 \end{bmatrix}$$

$$\mathsf{P}_2 \text{ policy} = \begin{cases} \begin{bmatrix} \frac{2}{3} \\ \frac{1}{3} \end{bmatrix} & \text{if IS} = \beta \\ \begin{bmatrix} 0 \\ 1 \end{bmatrix} & \text{if IS} = \xi \end{cases}$$

4. Replacement of information sets by the values of the games.

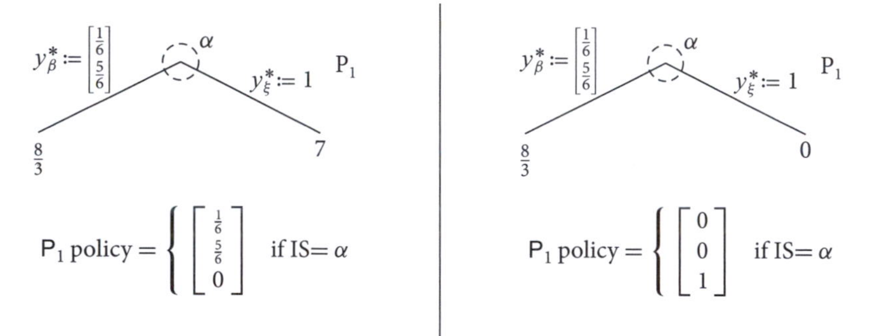

Figure 8.2 Recursive computation of behavioral saddle-point equilibria for two single-stage games following the procedure described in Section 8.5. As usual, for both games P_1 is the minimizer and P_2 the maximizer. We saw these games in Figure 7.4, where this procedure failed because, at the time, we were restricted to pure policies.

$$J\left(\{\gamma_{1\ldots K-1}, \boxed{\gamma_K^*}\}, \{\sigma_{1\ldots K-1}, \boxed{\sigma_K^*}\}\right)$$
$$\leq J\left(\{\gamma_{1\ldots K-1}, \boxed{\gamma_K}\}, \{\sigma_{1\ldots K-1}, \boxed{\sigma_K^*}\}\right), \quad \forall \gamma_K$$
$$J\left(\{\gamma_{1\ldots K-1}, \boxed{\gamma_K^*}\}, \{\sigma_{1\ldots K-1}, \boxed{\sigma_K^*}\}\right)$$
$$\geq J\left(\{\gamma_{1\ldots K-1}, \boxed{\gamma_K^*}\}, \{\sigma_{1\ldots K-1}, \boxed{\sigma_K}\}\right), \quad \forall \sigma_K$$

as required for a feedback behavioral saddle-point equilibrium.

5. Replace each information set for P_1 at the Kth stage by the value corresponding to the link selected by P_1.

   At this point we have a game with $K - 1$ stages, and we compute the sub-policies $(\gamma_{K-1}^*, \sigma_{K-1}^*)$ using the exact same procedure described above. This algorithm is iterated until all sub-policies have been obtained. Moreover, at every stage, we guarantee by construction that the conditions (8.3)–(8.4) for a feedback behavioral saddle-point equilibrium hold.

Attention! To reduce the number of stages in the game, we replace lower-level stages by the values obtained using the computed equilibrium sub-policies, which is precisely consistent with (8.3)–(8.4).

8.6 Mixed vs. Behavioral Order Interchangeability

We have seen that zero-sum feedback games in extensive form always have saddle-point equilibria in both mixed and behavioral policies. It turns out that an order interchangeability-like property exists for these two types on stochastic policies:

Lemma 8.2 (Mixed vs. behavioral order interchangeability). For every feedback game G in extensive form with a finite number of stages, if (γ_m^*, σ_m^*) is a mixed saddle-point equilibrium and (γ_b^*, σ_b^*) is a behavioral saddle-point equilibrium then

1. (γ_m^*, σ_b^*) is a saddle-point equilibrium for a game in which the action space of P_1 is the set of mixed polices and the action space of P_2 is the set of behavioral polices.
2. (γ_b^*, σ_m^*) is a saddle-point equilibrium for a game in which the action space of P_1 is the set of behavioral polices and the action space of P_2 is the set of mixed polices.

Moreover, all four equilibria have exactly the same value. □

The key consequence of this result is that, at least for zero-sum feedback games, there is little incentive in working in mixed policies since these are computationally more difficult and do not lead to benefits for the players.

8.7 Non-Feedback Games

Note. While there is no general recursive algorithm, there are algorithms for games with special information structures.

For non-feedback there is no known general recursive algorithm to compute saddle-point equilibria (pure, behavioral, or mixed). In fact, for non-feedback games, although mixed saddle-point equilibria always exist, behavioral saddle-point equilibria may not exist (cf. Figure 8.3).

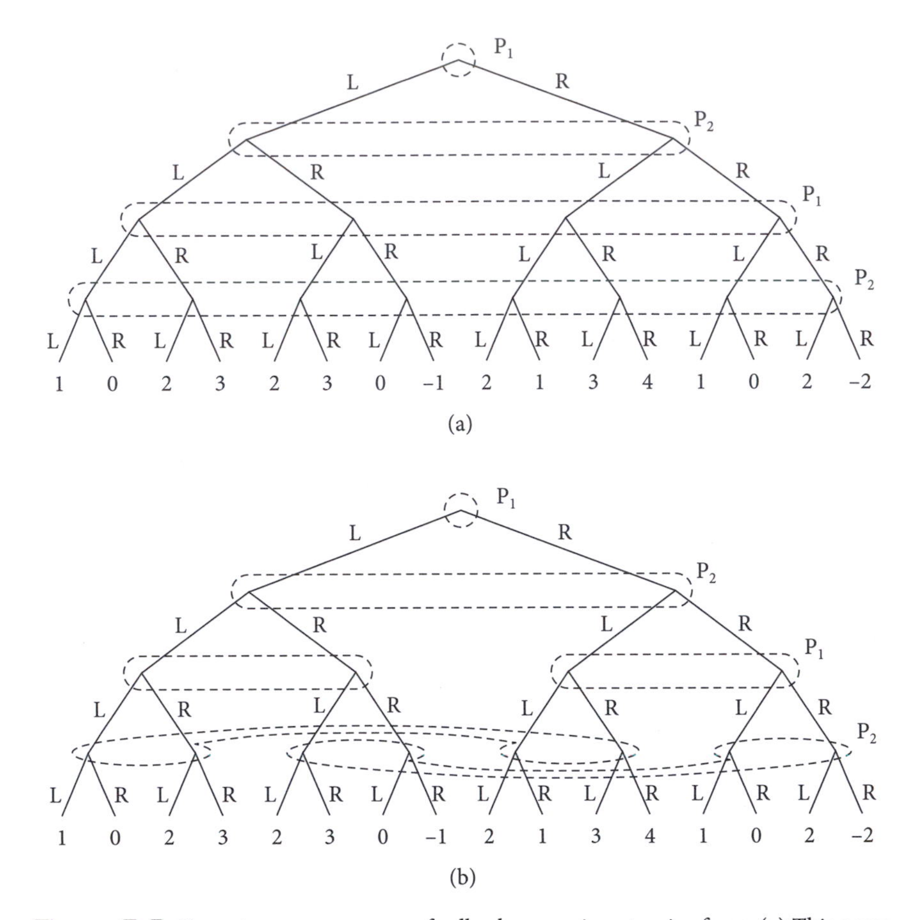

Figure 8.3 Two-stage zero-sum non-feedback games in extensive form. (a) This game does not have behavioral saddle-point equilibria [1, p. 52]. In this game each player knows nothing other than the current stage. In fact, both players do not even recall their previous choices. (b) This game has behavioral saddle-point equilibria [1, p. 55]. In this game, the players only know the current stage and their own previous decisions.

8.8 Practice Exercises

8.1. Consider the game in extensive form in Figure 8.1. Find mixed saddle-point equilibria for this game using policy domination and the graphical method.

Solution to Exercise 8.1. Suppose that we enumerate the policies available for P_1 and P_2 as follows:

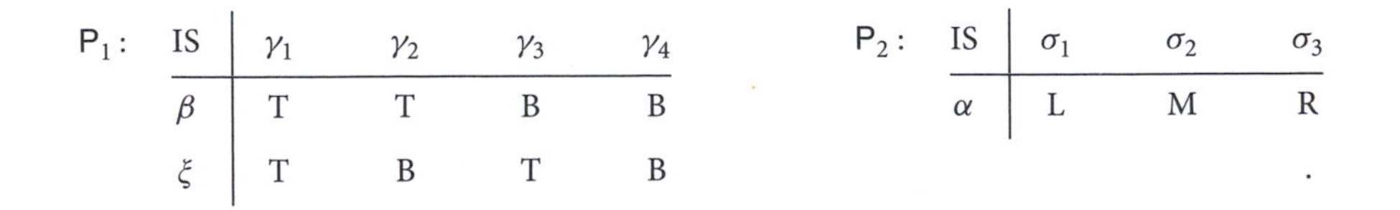

P_1:

IS	γ_1	γ_2	γ_3	γ_4
β	T	T	B	B
ξ	T	B	T	B

P_2:

IS	σ_1	σ_2	σ_3
α	L	M	R

.

In this case, we obtain

$$A_{\text{ext}} = \underbrace{\begin{bmatrix} 1 & 3 & 0 \\ 1 & 3 & 7 \\ 6 & 2 & 0 \\ 6 & 2 & 7 \end{bmatrix}}_{\substack{\mathsf{P}_2 \text{ policies} \\ (\sigma_1, \sigma_2, \sigma_3)}} \left.\vphantom{\begin{bmatrix} 1 \\ 1 \\ 6 \\ 6 \end{bmatrix}}\right\} \begin{matrix} \mathsf{P}_1 \text{ policies} \\ (\gamma_1, \gamma_2, \gamma_3, \gamma_4) \end{matrix} .$$

Using policy domination, this matrix can be reduced as follows:

$$A_{\text{ext}} \xrightarrow[\text{row 1 weakly dominates over 2}]{\text{row 3 weakly dominates over 4}} A^{\dagger}_{\text{ext}} := \begin{bmatrix} 1 & 3 & 0 \\ 6 & 2 & 0 \end{bmatrix} \xrightarrow[\text{col. 1 strictly dominates over 3}]{} A^{\ddagger}_{\text{ext}}$$
$$:= \begin{bmatrix} 1 & 3 \\ 6 & 2 \end{bmatrix}.$$

Using the graphical method we obtain

$$V_m(A_{\text{ext}}\ddagger) = \frac{8}{3}, \qquad y^* = \begin{bmatrix} \frac{2}{3} \\ \frac{1}{3} \end{bmatrix}, \qquad z^* = \begin{bmatrix} \frac{1}{6} \\ \frac{5}{6} \end{bmatrix},$$

which leads to the following value and mixed saddle-point equilibrium for the original game:

$$V_m(A_{\text{ext}}) = \frac{8}{3}, \qquad y^* = \begin{bmatrix} \frac{2}{3} \\ 0 \\ \frac{1}{3} \\ 0 \end{bmatrix}, \qquad z^* = \begin{bmatrix} \frac{1}{6} \\ \frac{5}{6} \\ 0 \end{bmatrix}.$$

8.2. Are there security policies for the game in Example 8.1, other than the ones given in equation (8.2)?

Solution to Exercise 8.2. Yes, we can take any convex combination of the policies for P_1 in equation (8.2): $y^* = [\frac{2}{3}\ 0\ \frac{\mu}{3}\ \frac{1-\mu}{3}]'$, $\forall \mu \in [0, 1]$. Moreover, we also lost other equilibria when we used weak domination to get the matrix $A^{\dagger}_{\text{ext}}$, from which this equilibrium was computed.

8.3. Compute pure or behavioral saddle-point equilibria for the games in extensive form depicted in Figure 8.4.

Solution to Exercise 8.3. See Figure 8.5.

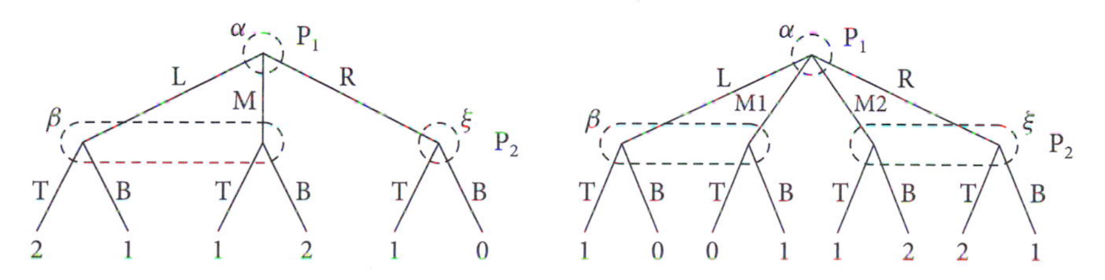

Figure 8.4 Single-stage games in extensive form for Exercise 8.3. For both games P_1 is the minimizer and P_2 the maximizer.

1. Initial games.

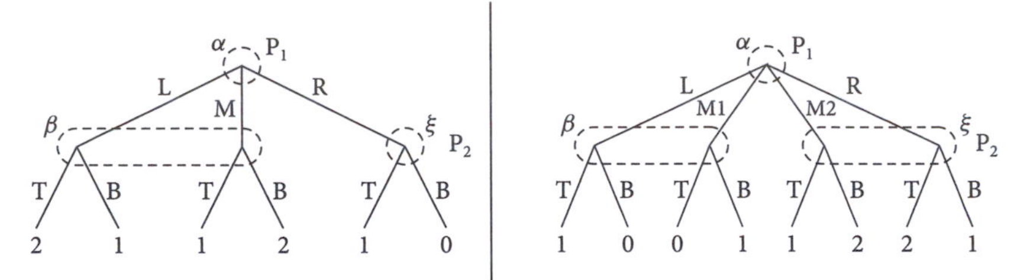

2. Construction of matrix games for the different information sets.

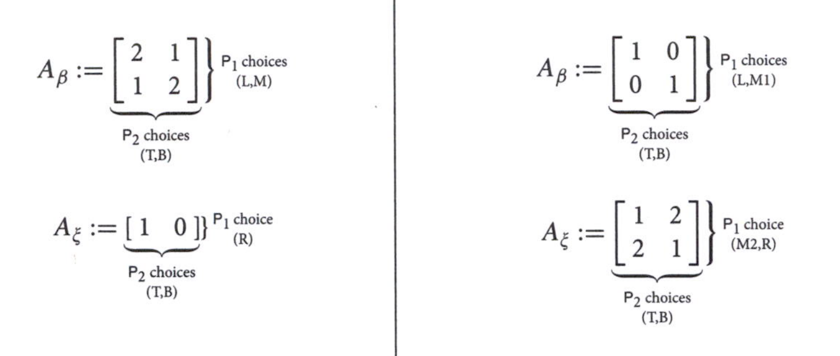

3. Solution of matrix games for the different information sets.

$$V_m(A_\beta) = \frac{3}{2},\ y^*_\beta := \begin{bmatrix} \frac{1}{2} \\ \frac{1}{2} \end{bmatrix},\ z^*_\beta := \begin{bmatrix} \frac{1}{2} \\ \frac{1}{2} \end{bmatrix} \qquad V_m(A_\beta) = \frac{1}{2},\ y^*_\beta := \begin{bmatrix} \frac{1}{2} \\ \frac{1}{2} \end{bmatrix},\ z^*_\beta := \begin{bmatrix} \frac{1}{2} \\ \frac{1}{2} \end{bmatrix}$$

$$V_m(A_\xi) = 1,\ y^*_\xi := 1,\quad z^*_\xi := \begin{bmatrix} 1 \\ 0 \end{bmatrix} \qquad V_m(A_\xi) = \frac{3}{2},\ y^*_\xi := \begin{bmatrix} \frac{1}{2} \\ \frac{1}{2} \end{bmatrix},\ z^*_\xi := \begin{bmatrix} \frac{1}{2} \\ \frac{1}{2} \end{bmatrix}$$

$$\mathsf{P}_2 \text{ policy} = \begin{cases} \begin{bmatrix} \frac{1}{2} \\ \frac{1}{2} \end{bmatrix} & \text{if IS} = \beta \\ \begin{bmatrix} 1 \\ 0 \end{bmatrix} & \text{if IS} = \xi \end{cases} \qquad \mathsf{P}_2 \text{ policy} = \left\{ \begin{bmatrix} \frac{1}{2} \\ \frac{1}{2} \end{bmatrix} \right. \ \text{if IS} = \beta \text{ or } \xi$$

4. Replacement of information sets by the values of the games.

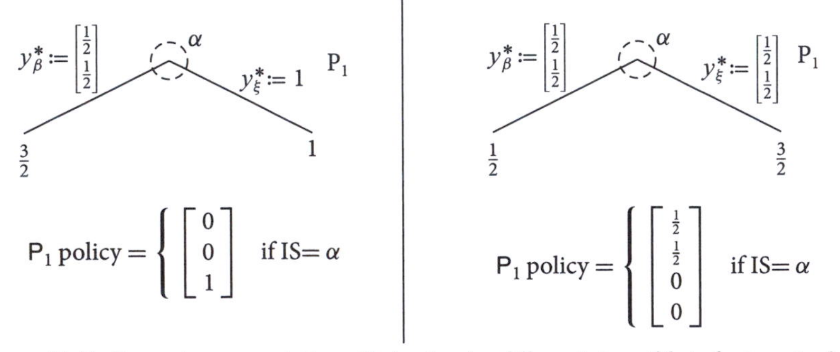

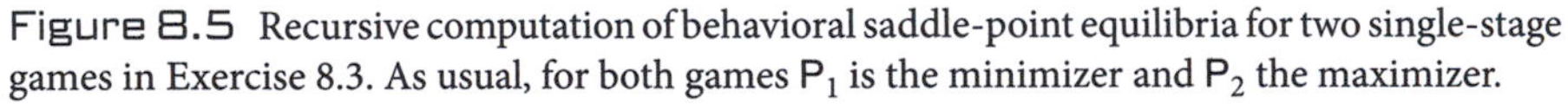

Figure 8.5 Recursive computation of behavioral saddle-point equilibria for two single-stage games in Exercise 8.3. As usual, for both games P_1 is the minimizer and P_2 the maximizer.

8.4. Consider the games in extensive form depicted in Figure 8.4 and behavioral policies of the form

	IS	1	2	$\cdots$	m	
P_1:	α	y_1^α	y_2^α	$\cdots$	y_m^α	$y_i^\alpha \geq 0, \forall i, \quad \sum_i y_i^\alpha = 1$
P_2:	IS	1	2			
	β	z_1^β	z_2^β			$z_1^\beta, z_2^\beta \geq 0, \quad z_1^\beta + z_2^\beta = 1$
	ξ	z_1^ξ	z_2^ξ			$z_1^\xi, z_2^\xi \geq 0, \quad z_1^\xi + z_2^\xi = 1.$

1. Show that the expected cost in both games can be written in the form

$$J(y^\alpha, z^\beta, z^\xi) := y^{\alpha\prime} A^\beta z^\beta + y^{\alpha\prime} A^\xi z^\xi, \tag{8.7}$$

where

$$y^\alpha := [\, y_1^\alpha y_2^\alpha \quad \cdots \quad y_m^\alpha \,]', \qquad z^\beta := \begin{bmatrix} z_1^\beta & z_2^\beta \end{bmatrix}', \qquad z^\xi := \begin{bmatrix} z_1^\xi & z_2^\xi \end{bmatrix}'.$$

2. Formulate the computation of the security policy for P_2 as a linear program. The matrices A^β and A^ξ in (8.7) should appear in this linear program.
3. Formulate the computation of the security policy for P_1 as a linear program. The matrices A^β and A^ξ in (8.7) should also appear in this linear program.
4. Use the two linear programs above to find behavioral saddle-point equilibria for the games in Figure 8.4 numerically using MATLAB®.

Solution to Exercise 8.4.

1. For the left hand side game in Figure 8.4, the expected cost is given by

$$\begin{aligned} J(y^\alpha, z^\beta, z^\xi) &= 2 \times y_1^\alpha z_1^\beta + 1 \times y_1^\alpha z_2^\beta + 1 \times y_2^\alpha z_1^\beta \\ &\quad + 2 \times y_2^\alpha z_2^\beta + 1 \times y_3^\alpha z_1^\xi + 0 \times y_3^\alpha z_2^\xi \\ &= y^{\alpha\prime} \begin{bmatrix} 2 & 1 \\ 1 & 2 \\ 0 & 0 \end{bmatrix} z^\beta + y^{\alpha\prime} \begin{bmatrix} 0 & 0 \\ 0 & 0 \\ 1 & 0 \end{bmatrix} z^\xi \end{aligned}$$

and for the right hand side game in Figure 8.4, the expected cost is given by

$$\begin{aligned} J(y^\alpha, z^\beta, z^\xi) &= 1 \times y_1^\alpha z_1^\beta + 0 \times y_1^\alpha z_2^\beta + 0 \times y_2^\alpha z_1^\beta + 1 \times y_2^\alpha z_2^\beta + 1 \times y_3^\alpha z_1^\xi \\ &\quad + 2 \times y_3^\alpha z_2^\xi + 2 \times y_4^\alpha z_1^\xi + 1 \times y_4^\alpha z_2^\xi \\ &= y^{\alpha\prime} \begin{bmatrix} 1 & 0 \\ 0 & 1 \\ 0 & 0 \\ 0 & 0 \end{bmatrix} z^\beta + y^{\alpha\prime} \begin{bmatrix} 0 & 0 \\ 0 & 0 \\ 1 & 2 \\ 2 & 1 \end{bmatrix} z^\xi. \end{aligned}$$

2. The security policy for P_2 arises out of the following minimax computation:

$$\bar{V}(G) := \max_{z^\beta, z^\xi \in \mathcal{Z} \subset \mathbb{R}^2} \min_{y^\alpha \in \mathcal{Y} \subset \mathbb{R}^3} y^{\alpha\prime} A^\beta z^\beta + y^{\alpha\prime} A^\xi z^\xi.$$

Focusing on the inner minimization, we conclude that

$$\begin{aligned}\min_{y^\alpha \in \mathcal{Y}} y^{\alpha\prime} A^\beta z^\beta + y^{\alpha\prime} A^\xi z^\xi &= \min_{i \in \{1,2,\ldots,m\}} \underbrace{(A^\beta z^\beta + A^\xi z^\xi)_i}_{i\text{th entry of } A^\beta z^\beta + A^\xi z^\xi} \\ &= \max\{v : v \le (A^\beta z^\beta + A^\xi z^\xi)_i, \ \ \forall i\}.\end{aligned}$$

Therefore

$$\begin{aligned}\bar{V}(G) &= \max_{z^\beta, z^\xi \in \mathcal{Z}} \max\{v : v \le (A^\beta z^\beta + A^\xi z^\xi)_i, \forall i\} \\ &= \underbrace{\begin{array}{ll} \text{maximum} & v \\ \text{subject to} & A^\beta z^\beta + A^\xi z^\xi \stackrel{.}{\ge} v\mathbf{1} \\ & \left.\begin{array}{r} z^\beta \stackrel{.}{\ge} 0 \\ \mathbf{1} z^\beta = 1 \end{array}\right\} (z^\beta \in \mathcal{Z}) \\ & \left.\begin{array}{r} z^\xi \stackrel{.}{\ge} 0 \\ \mathbf{1} z^\xi = 1 \end{array}\right\} (z^\xi \in \mathcal{Z}) \end{array}}_{\text{optimization over 4+1 parameters } (v, z_1^\beta, z_2^\beta, z_1^\xi, z_2^\xi)}.\end{aligned}$$

3. The security policy for P_1 arises out of the following minimax computation:

$$\bar{V}(G) := \min_{y^\alpha \in \mathcal{Y} \subset \mathbb{R}^3} \max_{z^\beta, z^\xi \in \mathcal{Z} \subset \mathbb{R}^2} y^{\alpha\prime} A^\beta z^\beta + y^{\alpha\prime} A^\xi z^\xi.$$

Since $y^{\alpha\prime} A^\beta z^\beta$ only depends on z^β and $y^{\alpha\prime} A^\xi z^\xi$ only depends on z^ξ, we have that

$$\begin{aligned}&\max_{z^\beta, z^\xi \in \mathcal{Z}} y^{\alpha\prime} A^\beta z^\beta + y^{\alpha\prime} A^\xi z^\xi \\ &\quad = \max_{z^\beta \in \mathcal{Z}} y^{\alpha\prime} A^\beta z^\beta + \max_{z^\xi \in \mathcal{Z}} y^{\alpha\prime} A^\xi z^\xi \\ &\quad = \max_{j \in \{1,2\}} \underbrace{(y^{\alpha\prime} A^\beta)_j}_{j\text{th entry of } y^{\alpha\prime} A^\beta} + \max_{j \in \{1,2\}} \underbrace{(y^{\alpha\prime} A^\xi)_j}_{j\text{th entry of } y^{\alpha\prime} A^\xi} \\ &\quad = \min\{v_1 : v_1 \ge (y^{\alpha\prime} A^\beta)_j, \forall j\} + \min\{v_2 : v_2 \ge (y^{\alpha\prime} A^\xi)_j, \forall j\} \\ &\quad = \min\{v_1 + v_2 : v_1 \ge (y^{\alpha\prime} A^\beta)_j, \forall j, \, v_2 \ge (y^{\alpha\prime} A^\xi)_j, \forall j\}.\end{aligned}$$

Therefore

$$\bar{V}(G) = \min_{y^\alpha \in \mathcal{Y}} \min\{v_1 + v_2 : v_1 \geq (y^{\alpha\prime} A^\beta)_j, \forall j, v_2 \geq (y^{\alpha\prime} A^\xi)_j, \forall j\}$$

$$= \underbrace{\begin{array}{ll} \text{minimum} & v_1 + v_2 \\ \text{subject to} & A^{\beta\prime} y^\alpha \dot{\leq} v_1 \mathbf{1} \\ & A^{\xi\prime} y^\alpha \dot{\leq} v_2 \mathbf{1} \\ & \left.\begin{array}{r} y^\alpha \dot{\geq} 0 \\ \mathbf{1} y^\alpha = 1 \end{array}\right\} (y \in \mathcal{Y}) \end{array}}_{\text{optimization over } m+2 \text{ parameters } (v_1, v_2, y_1^\alpha, y_2^\alpha, \dots, y_m^\alpha)}.$$

4. The following CVX code can be used to compute the behavioral saddle-point equilibria for the left hand side game in Figure 8.4:

```
Abeta=[2,1;1,2;0,0];
Axi  =[0,0;0,0;1,0];
% P1 security policy
cvx_begin
  variables v zbeta(size(Abeta,2)) zxi(size(Axi,2))
  maximize  v
  subject to
    Abeta*zbeta+Axi*zxi>=v
    zbeta>=0
    sum(zbeta)==1
    zxi>=0
    sum(zxi)==1
cvx_end
% P2 security policy
cvx_begin
  variables v1 v2 yalpha(size(Abeta,1))
  minimize  v1+v2
  subject to
    Abeta'*yalpha<=v1
    Axi'*yalpha<=v2
    yalpha>=0
    sum(yalpha)==1
cvx_end
```

which results in the following values

```
v  = 1.0000
zbeta   = 0.5000
          0.5000
zxi     = 1.0000
          0.0000
v1 = 1.4084e-09
v2 = 1.0000
yalpha = 0.0000
         0.0000
         1.0000
```

The CVX code to compute the behavioral saddle-point equilibria for the right hand side game in Figure 8.4 is similar to the one above, but starting with

```
Abeta=[1,0;0,1;0,0;0,0];
Axi   =[0,0;0,0;1,2;2,1];
  ⋮
```

which results in the following values

```
v   = 0.5000
zbeta   = 0.5000
          0.5000
zxi     = 0.5000
          0.5000
v1 = 0.5000
v2 = 2.9174e-10
yalpha = 0.5000
         0.5000
         0.0000
         0.0000
```

8.9 Additional Exercises

8.5. Prove Lemma 8.1. *Hint: follow the proof of Lemma 7.1.*

8.6. Consider a single-stage game in extensive form for which:

- P_1 is at the root of the tree with a single information set α and m actions $\{1, 2, \dots, m\}$.
- P_2 has ℓ information sets $\beta_1, \beta_2, \dots, \beta_\ell$ and a total of n actions $\{1, 2, \dots, n\}$.
- For each information set β_k, the set of actions of P_1 that lead to β_k is denoted by $\mathcal{A}_k$ and the set of actions of P_2 that exit β_k is denoted by $\mathcal{B}_k$.
- The outcome of the game when P_1 selects action $i \in \{1, 2, \dots, m\}$ and P_2 selects action $j \in \{1, 2, \dots, n\}$ is denoted by a_{ij}.

Suppose that the players use the following behavioral policies:

P_1:

IS	1	2	...	m
α	y_1^α	y_2^α	...	y_m^α

$y_i^\alpha \geq 0, \quad \forall i, \quad \sum_i y_i^\alpha = 1$

P_2:

IS	
β_1	$\{z_j^{\beta_1} : j \in \mathcal{B}_1\}$
β_2	$\{z_j^{\beta_2} : j \in \mathcal{B}_2\}$
$\vdots$	$\vdots$
β_ℓ	$\{z_j^{\beta_\ell} : j \in \mathcal{B}_\ell\}$

$$z_j^{\beta_1} \geq 0, \quad \forall j \in \mathcal{B}_1, \quad \sum_{j \in \mathcal{B}_1} z_j^{\beta_1} = 1$$
$$z_j^{\beta_2} \geq 0, \quad \forall j \in \mathcal{B}_2, \quad \sum_{j \in \mathcal{B}_2} z_j^{\beta_2} = 1$$
$$\vdots$$
$$z_j^{\beta_\ell} \geq 0, \quad \forall j \in \mathcal{B}_\ell, \quad \sum_{j \in \mathcal{B}_\ell} z_j^{\beta_\ell} = 1.$$

Write a formula for the expected outcome J for this game.

8.7. Consider the game in Exercise 8.6. For each player, construct a linear program to compute the player's behavioral saddle-point equilibrium policy.

PART III NON-ZERO-SUM GAMES

LECTURE 9

Two-Player Non-Zero-Sum Games

This lecture introduces non-zero-sum games for two players and defines several key concepts for such games.

9.1 Security Policies and Nash Equilibria

In this lecture we consider a two-player game G in which two players P_1 and P_2 are allowed to select policies within *action spaces* Γ_1 and Γ_2, respectively.

The key difference with respect to the games that we found so far is that we now assume that the game has a distinct *outcome* for each player. In particular, when

$$\begin{cases} \mathsf{P}_1 \text{ uses policy } \gamma \in \Gamma_1 \\ \mathsf{P}_2 \text{ uses policy } \sigma \in \Gamma_2 \end{cases}$$

we denote by

- $J_1(\gamma, \sigma)$ the *outcome of the game for player* P_1, and
- $J_2(\gamma, \sigma)$ the *outcome of the game for player* P_2.

Each player wants to *minimize* their own outcome, and does not care about the outcome of the other player.

Note. Since the objective is a minimization, the outcomes are generally called *costs*.

Notation. When $J_2(\gamma, \sigma) = -J_1(\gamma, \sigma), \forall \gamma, \sigma$ the game is called *zero-sum*. In such cases, as P_2 attempts to minimize $J_2(\gamma, \sigma)$ this player is also maximizing $-J_1(\gamma, \sigma)$ and we recover the types of games that we found before.

9.1.1 Security Levels and Policies

The definition of security policies for non-zero-sum games captures the notion of finding the policy that guarantees the least possible cost, assuming the worse possible choice (rational or not) by the other player.

Notation 1. The *infimum* of a set is its largest lower bound and the *supremum* of a set is its smallest upper bound. ⊳ p. 44

Notation. The security level for P_i only depends on J_i. This is emphasized by only using J_i as the argument for $\bar{V}_{\Gamma_1,\Gamma_2}(\cdot)$.

Definition 9.1 (Security policy). The *security level* for P_i, $i \in \{1, 2\}$ is defined by

$$\bar{V}_{\Gamma_1,\Gamma_2}(J_i) := \underbrace{\inf_{\gamma \in \Gamma_i}}_{\substack{\text{minimize cost assuming} \\ \text{worst choice by } \mathsf{P}_j}} \underbrace{\sup_{\sigma \in \Gamma_j, j \neq i}}_{\substack{\text{worst choice by } \mathsf{P}_j, \\ \text{from } \mathsf{P}_i\text{'s perspective}}} J_i(\gamma, \sigma)$$

Note. In general, security policies may not exist because the infimum may not be achieved by a policy in Γ_i.

and a *security policy* for P_i is any policy $\gamma^* \in \Gamma_i$ for which the infimum above is achieved, i.e.,

$$\bar{V}_{\Gamma_1,\Gamma_2}(J_i) := \inf_{\gamma \in \Gamma_i} \sup_{\sigma \in \Gamma_j, j \neq i} J_i(\gamma, \sigma) = \underbrace{\sup_{\sigma \in \Gamma_j, j \neq i} J_i(\gamma^*, \sigma)}_{\gamma^* \text{ achieves the infimum}}. \tag{9.1}$$

Notation. Equation (9.1) is often written as $\gamma^* \in \arg\min_\gamma \sup_\sigma J_i(\gamma, \sigma)$. The use of "$\in$" instead of "$=$" emphasizes that there may be several (or none) γ^* that achieve the infimum.

A pair of policies (γ^*, σ^*) is said to be a *minimax pair* if γ^* and σ^* are security policies for P_1 and P_2, respectively.

9.1.2 Nash Equilibria

The definition of equilibria for non-zero-sum games captures the notion of *no regret* in the sense that after knowing the choice made by the other player, each player finds that their own policy provided the lowest possible cost against the choice of the other player.

Note. These conditions are consistent with the Meta Definition 1.1.

Definition 9.2 (Nash equilibrium). A pair of policies $(\gamma^*, \sigma^*) \in \Gamma_1 \times \Gamma_2$ is called a *Nash equilibrium* if

$$J_1(\gamma^*, \sigma^*) \le J_1(\gamma, \sigma^*), \quad \forall \gamma \in \Gamma_1 \tag{9.2a}$$

$$J_2(\gamma^*, \sigma^*) \le J_2(\gamma^*, \sigma), \quad \forall \sigma \in \Gamma_2 \tag{9.2b}$$

and the pair $(J_1(\gamma^*, \sigma^*), J_2(\gamma^*, \sigma^*))$ is called the *Nash outcome of the game*.

Attention! For a *zero-sum game* in which $J_2(\gamma, \sigma) = -J_1(\gamma, \sigma)$, $\forall \gamma, \sigma$, (9.2b) becomes

$$-J_1(\gamma^*, \sigma^*) \le -J_1(\gamma^*, \sigma), \quad \forall \sigma \in \Gamma_2$$

and we can re-write (9.2a)–(9.2b) as

$$J_1(\gamma^*, \sigma) \le J_1(\gamma^*, \sigma^*) \le J_1(\gamma, \sigma^*), \quad \forall \gamma \in \Gamma_1 \ \sigma \in \Gamma_2,$$

Attention! However, it does not make sense to talk about saddle-point equilibria for a non-zero-sum game.

which is precisely consistent with the definition of a saddle-point equilibrium found in Section 4.5. One can therefore say that saddle-point equilibria are Nash equilibria for zero-sum games.

9.2 Bimatrix Games

Pure bimatrix games are played by two players, each selecting policies from finite *action spaces*:

- P_1 has available m actions: $\Gamma_1 := \{1, 2, \dots, m\}$
- P_2 has available n actions: $\Gamma_2 := \{1, 2, \dots, n\}$.

The *outcomes* for the players are quantified by two $m \times n$ matrices $A = [a_{ij}]$ and $B = [b_{ij}]$, one for player P_1 and the other for player P_2, respectively. When

Note. One can imagine that P_1 selects a row of A/B and P_2 selects a column of A/B.

$$\begin{cases} \mathsf{P}_1 \text{ selects action } i \in \Gamma_1 := \{1, 2, \dots, m\} \\ \mathsf{P}_2 \text{ selects action } j \in \Gamma_2 := \{1, 2, \dots, n\}, \end{cases}$$

we have that

$$\begin{cases} J_1 := a_{ij} \text{ is the outcome for } \mathsf{P}_1 \\ J_2 := b_{ij} \text{ is the outcome for } \mathsf{P}_2. \end{cases}$$

Both players want to *minimize* their respective outcomes J_1 and J_2. *Zero-sum games* can be viewed as a special case of bimatrix games, for which $B = -A$.

Note. When $B = -A$, as P_2 attempts to minimize $J_2 := b_{ij}$ she is also maximizing $-J_1 = -a_{ij}$.

For these action spaces, the resulting security levels, policies, and Nash equilibria defined in Section 9.1 are called *pure*.

Note. Since these action spaces are finite, security policies and minimax pairs always exist.

In what follows, we will see through examples of bimatrix games that non-zero-sum games exhibit important differences with respect to zero-sum games.

Example 9.1 (Prisoners' dilemma). Consider the bimatrix game defined by the following matrices

$$A = \underbrace{\begin{bmatrix} 2 & 30 \\ 0 & 8 \end{bmatrix}}_{\mathsf{P}_2 \text{ choices}} \Big\} \mathsf{P}_1 \text{ choices} \qquad B = \underbrace{\begin{bmatrix} 2 & 0 \\ 30 & 8 \end{bmatrix}}_{\mathsf{P}_2 \text{ choices}} \Big\} \mathsf{P}_1 \text{ choices}. \tag{9.3}$$

This game is called the *prisoners' dilemma* when one makes the following associations: P_1 and P_2 are two prisoners arrested for a minor crime, but suspected of having committed a serious crime. There is little evidence that incriminates them of the serious crime, so the prosecution's hope is that one of them incriminates the other. The prisoners thus have two options:

$$\begin{cases} \text{action 1 is "do not confess"} \\ \text{action 2 is "cooperate with the prosecution by testifying against other."} \end{cases}$$

The outcomes from (9.3) correspond to the number of years spent in jail resulting from their actions:

- if neither confesses, they are both convicted of the minor crime and spend 2 years in jail
- if they both cooperate they both get some deal, but still get 8 years for the serious crime

- if only one cooperates, the one that testifies is released but the other spends 30 years in prison.

Note. The two security policies correspond to confessing.

Notation. For the security level of the second player we use the notation $\bar{V}(B')$ for consistence with the previous notation for $\bar{V}(\cdot)$ used in zero-sum games, where the minimum is always taken over rows and the maximum over columns.

The security levels and policies for this game are

$$\bar{V}(A) = \min_i \max_j a_{ij} = 8, \qquad i^* = \arg\min_i \max_j a_{ij} = 2,$$

$$\bar{V}(B') = \min_j \max_i b_{ij} = 8, \qquad j^* = \arg\min_j \max_i b_{ij} = 2.$$

This game has a single Nash equilibrium which is also the minimax pair:

(2, 2) is a Nash equilibrium with outcome (8, 8).

Note. Because of this paradox, the prisoners' dilemma is the example most commonly used to illustrate the difference between cooperative and noncooperative solutions.

The apparent "paradox" is that the pair of policies (1, 1) would lead to the outcome (2, 2), which is a strict improvement for both players. However, within the context of noncooperative games, there is no paradox since this solution is not robust because either player can profit significantly from deviating from it so its "implementation" requires cooperation and mutual trust. □

Note. Games in which both players have the same outcome are called *identical interests games* and will be further discussed in Lecture 12. ⊳ p. 133.

Attention! Bimatrix games are meaningful even when both players have the same objective, i.e., when $A = B$. However, we are still interested in *noncooperative* solutions to such games, which means that we are mostly interested in solutions that are reached by the players without negotiation or faith/trust between them. This sometimes leads to apparent paradoxes. For example, we shall see shortly that the bimatrix game defined by

$$A = B = \underbrace{\begin{bmatrix} 0 & 2 \\ 2 & 1 \end{bmatrix}}_{\text{P}_2 \text{ choices}} \Big\} \text{P}_1 \text{ choices}$$

has two noncooperative equilibria in the sense of "no regret": one of them (1, 1) is optimal in a cooperative sense, but the other one (2, 2) is definitely not the cooperative solution.

9.3 Admissible Nash Equilibria

Example 9.2. Consider the bimatrix game defined by the following matrices

$$A = \underbrace{\begin{bmatrix} 1 & 0 \\ 2 & -1 \end{bmatrix}}_{\text{P}_2 \text{ choices}} \Big\} \text{P}_1 \text{ choices} \qquad B = \underbrace{\begin{bmatrix} 2 & 3 \\ 1 & 0 \end{bmatrix}}_{\text{P}_2 \text{ choices}} \Big\} \text{P}_1 \text{ choices}.$$

The security levels and policies for this game are

$$\bar{V}(A) = \min_i \max_j a_{ij} = 1, \qquad i^* = \arg\min_i \max_j a_{ij} = 1,$$

$$\bar{V}(B') = \min_j \max_i b_{ij} = 2, \qquad j^* = \arg\min_j \max_i b_{ij} = 1.$$

In spite of the fact that this game has unique security policies, it has two Nash equilibria

$$(1, 1) \text{ is a Nash equilibrium with outcome } (1, 2) \tag{9.4}$$

$$(2, 2) \text{ is a Nash equilibrium with outcome } (-1, 0). \tag{9.5}$$

□

This example illustrates a few key differences between bimatrix games and zero-sum matrix games:

- Bimatrix games may have several Nash equilibria (like zero-sum games that may have several saddle-point equilibria). However,
- Nash equilibria are not always security policies (*unlike* zero-sum games for which saddle-point equilibria are always security policies).
- Different equilibria to bimatrix games may have different outcomes (*unlike* zero-sum games for which saddle-point equilibria always have the same value).

Note. This definition can be formulated for the general games in Section 9.1.

The fact that different equilibria may have different values, motivates the following definition:

Note. Both players do no worse with $(\bar{\gamma}^*, \bar{\sigma}^*)$ and at least one of them does strictly better.

Example. In Example 9.2 only (9.5) is an admissible Nash equilibrium.

Definition 9.3 (Admissible Nash equilibria). A Nash equilibrium $(\gamma^*, \sigma^*) \in \Gamma_1 \times \Gamma_2$ is said to be *admissible* if there is no "better" Nash equilibrium in the sense that there is no other Nash equilibrium $(\bar{\gamma}^*, \bar{\sigma}^*) \in \Gamma_1 \times \Gamma_2$ such that

$$J_1(\bar{\gamma}^*, \bar{\sigma}^*) \leq J_1(\gamma^*, \sigma^*), \qquad J_2(\bar{\gamma}^*, \bar{\sigma}^*) \leq J_2(\gamma^*, \sigma^*),$$

with at least one of these inequalities strict.

Notation. This game is called the *battle of the sexes*. ▷ p. 110

Example 9.3 (Battle of the sexes). Consider the bimatrix game defined by the following matrices

$$A = \underbrace{\begin{bmatrix} -2 & 1 \\ 0 & -1 \end{bmatrix}}_{\text{P}_2 \text{ choices}} \Big\} \text{P}_1 \text{ choices} \qquad B = \underbrace{\begin{bmatrix} -1 & 3 \\ 2 & -2 \end{bmatrix}}_{\text{P}_2 \text{ choices}} \Big\} \text{P}_1 \text{ choices}. \tag{9.6}$$

Note. Security levels can be very conservative, in fact often much worse than any of the Nash equilibria.

The security levels and policies for this game are

$$\bar{V}(A) = \min_i \max_j a_{ij} = 0, \qquad i^* = \arg\min_i \max_j a_{ij} = 2,$$

$$\bar{V}(B') = \min_j \max_i b_{ij} = 2, \qquad j^* = \arg\min_j \max_i b_{ij} = 1.$$

Attention! The minimax pair (1, 2) is *not* a Nash equilibrium.

This game has two Nash equilibria

$$(1, 1) \text{ is a Nash equilibrium with outcome } (-2, -1)$$

$$(2, 2) \text{ is a Nash equilibrium with outcome } (-1, -2)$$

and both equilibria are admissible since none is "better" than the other. □

This example illustrates two additional important properties of bimatrix games

- Bimatrix games may have several admissible Nash equilibria.
- Nash equilibria are not interchangeable, in the sense that (γ_1^*, σ_1^*) and (γ_2^*, σ_2^*) may both be Nash equilibria, but (γ_1^*, σ_2^*) and (γ_2^*, σ_1^*) may not be.

Note. However, we will see shortly that in some cases we may have interchangeability.

Example. In Example 9.3, neither (1,2) nor (2,1) are Nash equilibria since in both cases, both players regret their choices and, after knowing the other player's decision, they wish they had done things differently.

Notation 9.3 (Battle of the sexes). The game in Example 9.3 is called the *battle of the sexes* when one makes the following associations:

$$\begin{cases} \mathsf{P}_1 \text{ is the husband/boyfriend} \\ \mathsf{P}_2 \text{ is the wife/girlfriend} \end{cases} \quad \begin{cases} \text{action 1 corresponds to “going to a baby shower”} \\ \text{action 2 corresponds to “going to a football game.”} \end{cases}$$

Under the outcomes provided by (9.6):

- they have the most fun if they go together, but
- the husband prefers the baby shower,
- the wife prefers the football game.

Note. It is not uncommon to find less politically correct versions of this game.

Attention! Do not use results from noncooperative game theory to resolve your dating problems!!!

Although some form of alternation (perhaps stochastically by tossing a coin) seems like a reasonable solution to the battle of the sexes, it turns out that there is no such admissible Nash equilibrium to this particular version of the battle of the sexes (which we will be able to verify as soon as we introduce mixed policies for these games). □

9.4 Mixed Policies

Bimatrix games also admit *mixed policies*, which consist of choosing a probability distribution to select actions. As in zero-sum matrix games, the idea behind mixed policies is that the players select their actions randomly according to previously selected probability distributions.

Consider a game specified by two $m \times n$ matrices A and B that determine the outcomes for P_1 and P_2, respectively:

- A *mixed policy for* P_1 is a set of numbers

$$\{y_1, y_2, \ldots, y_m\}, \quad \sum_{i=1}^{m} y_i = 1 \quad y_i \geq 0, \quad \forall i \in \{1, 2, \ldots, m\},$$

  with the understanding that y_i is the probability that P_1 uses to select the action $i \in \{1, 2, \ldots, m\}$.

- A *mixed policy for* P_2 is a set of numbers

$$\{z_1, z_2, \ldots, z_n\}, \quad \sum_{j=1}^{n} z_j = 1 \quad z_j \geq 0, \quad \forall j \in \{1, 2, \ldots, n\},$$

  with the understanding that z_j is the probability that P_2 uses to select action $j \in \{1, 2, \ldots, n\}$.

It is assumed that the random selections by both players are done statistically independently and the players P_1 and P_2 will try to minimize their *expected outcomes*, which are respectively given by

Note. See Section 4.1 on how to derive these expressions.

$$J_1 = \sum_{i,j} a_{ij} y_i z_j = y' A z, \qquad J_2 = \sum_{i,j} b_{ij} y_i z_j = y' B z,$$

where $y := [\, y_1 \;\; y_2 \;\; \cdots \;\; y_m \,]'$ and $z := [\, z_1 \;\; z_2 \;\; \cdots \;\; z_n \,]'$.

Having defined mixed policies for a bimatrix game, we can use the concepts of security levels, security policies, and Nash equilibria introduced in Section 9.1 for general games, with the understanding that:

1. the *action spaces* are the sets $\mathcal{Y}$ and $\mathcal{Z}$ of all mixed policies for players P_1 and P_2, respectively; and
2. for a particular pair of mixed policies $y \in \mathcal{Y}$, $z \in \mathcal{Z}$ for P_1 and P_2, respectively,
   - $J_1(y, z) := y' A z$ is the outcome for P_1,
   - $J_2(y, z) := y' B z$ is the outcome for P_2.

In particular, this leads to the following definition of saddle-point equilibrium:

Note. One can also define average security levels and mixed policies. Mixed security policies never lead to worse outcomes and often result in strictly lower costs (cf. Exercise 9.3). This means that a player can always self protect better (or at least no worse) with mixed policies (even when the other player is also allowed to use mixed policies).

Definition 9.4 (Mixed Nash equilibrium). A pair of policies $(y^*, z^*) \in \mathcal{Y} \times \mathcal{Z}$ is called a *mixed Nash equilibrium* if

$$y^{*\prime} A z^* \leq y' A z^*, \quad \forall y \in \mathcal{Y}, \qquad y^{*\prime} B z^* \leq y^{*\prime} B z, \quad \forall z \in \mathcal{Z}$$

and $(y^{*\prime} A z^*, y^{*\prime} B z^*)$ is called the *mixed Nash outcome of the game.*

As in zero-sum games, the introduction of mixed policies enlarged the action spaces for both players to the point that Nash equilibria now always exist:

Theorem 9.1 (Nash). Every bimatrix game has at least one mixed Nash equilibrium. □

Note. A proof of this result, due to Nash [10], can be constructed from Brouwer's Fixed-point Theorem.

9.5 Best-Response Equivalent Games and Order Interchangeability

Consider two general two-player games G and H with the same *action spaces* Γ_1 and Γ_2 but different outcomes. Specifically, for the same pair of policies $\gamma \in \Gamma_1, \sigma \in \Gamma_2$,

- G has *outcomes* $G_1(\gamma, \sigma)$ and $G_2(\gamma, \sigma)$ for the players P_1 and P_2, respectively,
- H has *outcomes* $H_1(\gamma, \sigma)$ and $H_2(\gamma, \sigma)$ for players P_1 and P_2, respectively.

Definition 9.5 (Best-response equivalent). The games G and H are said to be *best-response equivalent* if they have the same set of Nash equilibria, i.e., a pair of policies (γ, σ) is a Nash equilibrium for G if and only if it is a Nash equilibrium for H.

Best-response equivalence allows us to characterize a class of games for which we have order interchangeability for Nash equilibria:

Attention! In general, the converse of this result is not true. In particular, the multiple Nash equilibria of a two-player game may be interchangeable, but the game may not be best-response equivalent to a zero-sum game.

Proposition 9.1 (Order interchangeability). The Nash equilibria of the matrix game G are interchangeable if G is best-response equivalent to a zero-sum game H that is zero-sum. □

Proof of Proposition 9.1. This proposition is a straightforward consequence of the facts that (i) zero-sum games enjoy the order interchangeability property that we found in Proposition 4.2 (item refpr:general-order) and (ii) if two games are best-response equivalent they have the same Nash equilibria.

Indeed, if (γ_1^*, σ_1^*) and (γ_2^*, σ_2^*) are both Nash equilibria for the game G, then because of best-response equivalence (γ_1^*, σ_1^*) and (γ_2^*, σ_2^*) are also Nash equilibria for the game H. But since H is zero-sum, we conclude from Proposition 4.2 that (γ_1^*, σ_2^*) and (γ_1^*, σ_2^*) are also Nash equilibria for H. Again, because of best-response equivalence we conclude that (γ_1^*, σ_2^*) and (γ_1^*, σ_2^*) must be Nash equilibria for G. ■

It turns out that it may be possible to show that two games are best-response equivalent simply by examining the functions that define their outcomes, without explicitly computing their Nash equilibria.

Notation 2. A function f is monotone strictly increasing if $x > y \Rightarrow f(x) > f(y), \forall x, y$.
▷ p. 113

Lemma 9.1. Suppose that there exist two monotone strictly increasing scalar functions $\alpha : \mathbb{R} \to \mathbb{R}$ and $\beta : \mathbb{R} \to \mathbb{R}$ such that

$$H_1(\gamma, \sigma) = \alpha(G_1(\gamma, \sigma)), \quad H_2(\gamma, \sigma) = \beta(G_2(\gamma, \sigma)), \quad \forall \gamma \in \Gamma_1, \sigma \in \Gamma_2, \quad (9.7)$$

then G and H are best-response equivalent. □

Note. The condition provided in Lemma 9.1 is sufficient for best-response equivalence, but it is not necessary, in the sense that there may be games that are best-response equivalent, but for which the outcomes are not related by expressions of the form (9.7).

Proof of Lemma 9.1. We start by showing that if (γ^*, σ^*) is a Nash equilibrium of G then it is also a Nash equilibrium of H. To this effect, we assume that we are given a Nash equilibrium (γ^*, σ^*) of G, for which

$$G_1(\gamma^*, \sigma^*) \le G_1(\gamma, \sigma^*), \quad \forall \gamma, \qquad G_2(\gamma^*, \sigma^*) \le G_2(\gamma^*, \sigma), \quad \forall \sigma.$$

Applying the monotone functions α and β to both sides of the left and right-hand side inequalities, respectively, we obtain

Note. This step only requires α and β to be monotone nondecreasing, but not strictly increasing.

$$\alpha(G_1(\gamma^*, \sigma^*)) \le \alpha(G_1(\gamma, \sigma^*)), \quad \forall \gamma, \qquad \beta(G_2(\gamma^*, \sigma^*)) \le \beta(G_2(\gamma^*, \sigma)), \quad \forall \sigma.$$

From this and (9.7), we conclude that

$$H_1(\gamma^*, \sigma^*) \le H_1(\gamma, \sigma^*), \quad \forall \gamma, \qquad H_2(\gamma^*, \sigma^*) \le H_2(\gamma^*, \sigma), \quad \forall \sigma,$$

which confirms that (γ^*, σ^*) is indeed a Nash equilibrium of H.

Suppose now that we are given a Nash equilibrium (γ^*, σ^*) of H, for which

$$H_1(\gamma^*, \sigma^*) \le H_1(\gamma, \sigma^*), \quad \forall \gamma, \qquad H_2(\gamma^*, \sigma^*) \le H_2(\gamma^*, \sigma), \quad \forall \sigma.$$

If we now combine this with (9.7), we obtain

$$\alpha(G_1(\gamma^*, \sigma^*)) \le \alpha(G_1(\gamma, \sigma^*)), \quad \forall \gamma, \qquad \beta(G_2(\gamma^*, \sigma^*)) \le \beta(G_2(\gamma^*, \sigma)), \quad \forall \sigma.$$

But since the functions α and β are monotone strictly increasing this implies that

$$G_1(\gamma^*, \sigma^*) \le G_1(\gamma, \sigma^*), \quad \forall\gamma, \qquad G_2(\gamma^*, \sigma^*) \le G_2(\gamma^*, \sigma), \quad \forall\sigma,$$

which confirms that (γ^*, σ^*) is indeed a Nash equilibrium of G. ■

Note. This step uses Note 10 (p. 113) and does require *strict* monotonicity since otherwise, e.g., equality of $\alpha(G_1(\gamma^*, \sigma^*))$ and $\alpha(G_1(\gamma, \sigma^*))$ would tell us little regarding how $G_1(\gamma^*, \sigma^*)$ compares to $G_1(\gamma, \sigma^*)$.

Attention! Given a two-player game G one can try to find monotonically strictly increasing functions $\alpha, \beta : \mathbb{R} \to \mathbb{R}$ such that (9.7) holds with $H_1(\gamma, \sigma) = -H_2(\gamma, \sigma)$, $\forall\gamma, \sigma$, which would allow us to conclude that the Nash equilibria of G are interchangeable and all Nash outcomes are equal to each other.

Note. See Exercise 9.4. ▷ p. 116

Specializing this to a bimatrix game G defined by a pair of $m \times n$ matrices A and B and pure policies, this amounts to finding monotonically strictly increasing functions $\alpha, \beta : \mathbb{R} \to \mathbb{R}$ such that

$$\alpha(a_{ij}) = -\beta(b_{ij}), \quad \forall i, j. \tag{9.8}$$

By restricting our search, e.g., to polynomial functions of the type

$$\alpha(s) := \sum_{k=1}^{\ell} a_k s^k, \qquad \beta(s) := \sum_{k=1}^{\ell} b_k s^k,$$

the conditions in (9.8) lead to linear equations on the polynomial coefficients, which are easy to solve. However, one would still need to verify the monotonicity of the polynomials so obtained (over the range of possible game outcomes).

Note. If the functions $\alpha, \beta : \mathbb{R} \to \mathbb{R}$ are strictly monotone increasing and affine, then (9.8) would still suffice to conclude equivalence to a zero-sum game under mixed policies. See Exercise 9.2. ▷ p. 115

Notation 2 (Monotone function). A function $f : \mathbb{R} \to \mathbb{R}$ is said to be *monotone nondecreasing* if

$$x \ge y \;\Rightarrow\; f(x) \ge f(y), \quad \forall x, y \in \mathbb{R},$$

and it is said to be *monotone strictly increasing* if

$$x > y \;\Rightarrow\; f(x) > f(y), \quad \forall x, y \in \mathbb{R}.$$ □

Note 10. Strict monotonicity is also equivalent to say that $f(x) \le f(y) \Rightarrow x \le y, \ \forall x, y$.

Attention! The lack of interchangeability is an "unpleasant" possibility in non-zero-sum games and leads to the following hierarchy of two-player games:

1. Games with a single Nash equilibrium or with multiple but interchangeable Nash equilibria with equal values are the most "predictable" for noncooperative rational players.

   This class of games includes all zero-sum games and the prisoners' dilemma.
2. Games with a single admissible Nash equilibrium or with multiple but interchangeable admissible Nash equilibria with equal values are still fairly predictable for noncooperative rational players.

   This class of games includes the bimatrix game considered in Example 9.2 or the one defined by $A = B = \begin{bmatrix} 0 & 2 \\ 2 & 1 \end{bmatrix}$ with a single admissible Nash equilibrium (1, 1). Note that (2, 2) is also a Nash equilibrium, but it is not admissible.

3. In games with multiple admissible Nash equilibria that are interchangeable but have different values, noncooperative rational players will likely end up in a Nash equilibrium, but it will generally be difficult to predict which.
4. Games with multiple admissible Nash equilibria that are not interchangeable are problematic since it is unclear whether or not the players will find a common equilibrium.

   This class of games includes the battle of the sexes or the bimatrix game defined by $A = B = \begin{bmatrix} 0 & 1 \\ 1 & 0 \end{bmatrix}$ with two admissible but non-interchangeable Nash equilibria (1, 1) and (2, 2) with the same value (0, 0).

   When played repeatedly, these games can lead to persistent oscillations in the policies used by the players as they may try to constantly adjust to the most recent policy used by the other player.

So one may ask what options do we have for the latter types of games in a noncooperative setting in which one should not rely on negotiation/trust between players:

1. The players may simply use security policies, leading to minimax solutions. Such solutions are often costly for both players and therefore not efficient.
2. When possible, the reward structure of the game should be changed to avoid inefficient solutions and policy oscillations in repeated games.

   It is often possible to "reshape" the reward structure of a game in economics (and engineering) through pricing, taxation, or other incentives/deterrents.

9.6 Practice Exercises

9.1 (Order interchangeability for Nash). Consider two Nash equilibria (γ_1^*, σ_1^*) and (γ_2^*, σ_2^*) for a two-player game. Show that if these two equilibria are interchangeable in the sense that (γ_1^*, σ_2^*) and (γ_2^*, σ_1^*) are also Nash equilibria, then

$$G_1(\gamma_1^*, \sigma_1^*) = G_1(\gamma_2^*, \sigma_1^*), \qquad G_1(\gamma_2^*, \sigma_2^*) = G_1(\gamma_1^*, \sigma_2^*),$$
$$G_2(\gamma_1^*, \sigma_1^*) = G_2(\gamma_1^*, \sigma_2^*), \qquad G_2(\gamma_2^*, \sigma_2^*) = G_2(\gamma_2^*, \sigma_1^*).$$

Solution to Exercise 9.1. Since (γ_1^*, σ_1^*) is a Nash equilibrium, we must have

$$G_1(\gamma_1^*, \sigma_1^*) \leq G_1(\gamma_2^*, \sigma_1^*), \qquad G_2(\gamma_1^*, \sigma_1^*) \leq G_2(\gamma_1^*, \sigma_2^*)$$

but since (γ_2^*, σ_1^*) and (γ_1^*, σ_2^*) are also Nash equilibria we must also have that

$$G_1(\gamma_2^*, \sigma_1^*) \leq G_1(\gamma_1^*, \sigma_1^*), \qquad G_2(\gamma_1^*, \sigma_2^*) \leq G_2(\gamma_1^*, \sigma_1^*),$$

therefore we actually have

$$G_1(\gamma_1^*, \sigma_1^*) = G_1(\gamma_2^*, \sigma_1^*), \qquad G_2(\gamma_1^*, \sigma_1^*) = G_2(\gamma_1^*, \sigma_2^*).$$

Similarly, using the facts that (γ_2^*, σ_2^*), (γ_1^*, σ_2^*), and (γ_2^*, σ_1^*) are all Nash equilibria, we can also conclude that

$$G_1(\gamma_2^*, \sigma_2^*) = G_1(\gamma_1^*, \sigma_2^*), \qquad G_2(\gamma_2^*, \sigma_2^*) = G_2(\gamma_2^*, \sigma_1^*),$$

which concludes the proof.

9.2 (Best-response equivalence for bimatrix games). Consider two bimatrix games: the game G is defined by the $m \times n$ matrices $A := [a_{ij}]$, $B := [b_{ij}]$ and the game H is defined by the matrices $\bar{A} := [\bar{a}_{ij}]$, $\bar{B} := [\bar{b}_{ij}]$ with

$$\bar{a}_{ij} = \alpha(a_{ij}), \quad \bar{b}_{ij} = \beta(b_{ij}), \quad \forall i \in \{1, 2, \ldots, m\}, \; j \in \{1, 2, \ldots, n\} \tag{9.9}$$

for some monotone strictly increasing scalar functions $\alpha : \mathbb{R} \to \mathbb{R}$ and $\beta : \mathbb{R} \to \mathbb{R}$.

1. Show that G and H are best-response equivalent in pure policies.
2. Show that if the functions α and β are affine, then G and H are also best-response equivalent in mixed policies.

   A function $f : \mathbb{R} \to \mathbb{R}$ is called *affine* if the function $\bar{f}(s) := f(s) - f(0)$ is linear, i.e., if $\bar{f}(\sum_i c_i x_i) = \sum_i c_i \bar{f}(x_i)$ or equivalently if $f(\sum_i c_i x_i) - f(0) = \sum_i c_i (f(x_i) - f(0))$.

Solution to Exercise 9.2.

1. In pure policies the action spaces are $\Gamma_1 := \{1, 2, \ldots, m\}$ and $\Gamma_2 := \{1, 2, \ldots, n\}$. In this case, given two pure policies $i \in \Gamma_1$ and $j \in \Gamma_2$ the outcomes for the two games are

$$G_1(i, j) = a_{ij}, \qquad G_2(i, j) = b_{ij},$$
$$H_1(i, j) = \bar{a}_{ij}, \qquad H_2(i, j) = \bar{b}_{ij},$$

   and therefore (9.9) is precisely the same as (9.7).

2. In mixed policies the action spaces are simplexes $\Gamma_i := \mathcal{Y}$ and $\Gamma_2 := \mathcal{Z}$. In this case, given two mixed policies $y \in \Gamma_1$ and $z \in \Gamma_2$ the outcomes for the two games are

$$G_1(y, z) = y'Az, \qquad G_2(y, z) = y'Bz,$$
$$H_1(y, z) = y'\bar{A}z, \qquad H_2(y, z) = y'\bar{B}z$$

   and therefore

$$\alpha(G_1(y, z)) = \alpha(y'Az) = \alpha\left(\sum_{i=1}^{m}\sum_{j=1}^{n} y_i a_{ij} z_j\right).$$

Since $\alpha(\cdot)$ is affine, $\alpha(\cdot) - \alpha(0)$ is a linear function and therefore

$$\begin{aligned}
\alpha(G_1(y, z)) - \alpha(0) &= \alpha\left(\sum_{i=1}^{m}\sum_{j=1}^{n} y_i a_{ij} z_j\right) - \alpha(0)\\
&= \sum_{i=1}^{m}\sum_{j=1}^{n} y_i z_j(\alpha(a_{ij}) - \alpha(0))\\
&= \sum_{i=1}^{m}\sum_{j=1}^{n} y_i z_j \alpha(a_{ij}) - \alpha(0)\sum_{i=1}^{m}\sum_{j=1}^{n} y_i z_j\\
&= y'\bar{A}z - \alpha(0) = H_1(y, z) - \alpha(0),
\end{aligned}$$

from which we obtain

$$H_1(y, z) = \alpha(G_1(y, z)), \quad \forall y \in \mathcal{Y}.$$

On the other hand, since $\beta(\cdot)$ is affine, $\beta(\cdot) - \beta(0)$ is linear and we conclude that

$$\begin{aligned}
\beta(G_2(y, z)) - \beta(0) &= \beta\left(\sum_{i=1}^{m}\sum_{j=1}^{n} y_i b_{ij} z_j\right) - \beta(0)\\
&= \sum_{i=1}^{m}\sum_{j=1}^{n} y_i z_j(\beta(b_{ij}) - \beta(0))\\
&= \sum_{i=1}^{m}\sum_{j=1}^{n} y_i z_j \beta(b_{ij}) - \beta(0)\sum_{i=1}^{m}\sum_{j=1}^{n} y_i z_j\\
&= y'\bar{B}z - \beta(0) = H_2(y, z) - \beta(0),
\end{aligned}$$

from which we obtain

$$H_2(y, z) = \beta(G_2(y, z)), \quad \forall z \in \mathcal{Z}.$$

This shows that (9.7) also holds in this case.

9.7 Additional Exercises

9.3. Show that for any bimatrix game, the mixed security level for a given player P_i is always smaller (i.e., better) than or equal to the pure security level for the same player P_i.

9.4 (Best response equivalence with equal outcomes). Suppose that

1. G is a game with outcomes $G_1(\gamma, \sigma)$ and $G_2(\gamma, \sigma)$ for the players P_1 and P_2, respectively;

2. H a zero-sum game with outcomes $H_1(\gamma, \sigma)$ and $H_2(\gamma, \sigma) = -H_1(\gamma, \sigma)$ for players P_1 and P_2, respectively; and
3. there exist two monotone strictly increasing scalar functions $\alpha : \mathbb{R} \to \mathbb{R}$ and $\beta : \mathbb{R} \to \mathbb{R}$ such that

$$H_1(\gamma, \sigma) = \alpha(G_1(\gamma, \sigma)), \quad H_2(\gamma, \sigma) = \beta(G_2(\gamma, \sigma)), \quad \forall \gamma \in \Gamma_1, \ \ \sigma \in \Gamma_2. \tag{9.10}$$

Show that G and H are best-response equivalent and all Nash outcomes of G are equal to each other.

Hint: The fact that G and H are best-response equivalent has already been proved in Lemma 9.1, so the task is to show that all Nash outcomes of G are equal to each other. Use the fact that all Nash equilibria of a zero-sum game are equal to each other (cf. Proposition 4.2).

LECTURE 10

Computation of Nash Equilibria for Bimatrix Games

This lecture addresses the computation of mixed Nash equilibria for bimatrix games.

10.1 Completely Mixed Nash Equilibria

Consider a different version of the battle of the sexes game introduced in Example 9.3, that is now defined by the following matrices

Note. Recall that action 1 corresponds to going to the baby shower and action 2 to the football game. In this version no one really wants to go to the football game alone (cost of 3), but going to the baby shower alone is a little better (cost of 0).

$$A = \underbrace{\begin{bmatrix} -2 & 0 \\ 3 & -1 \end{bmatrix}}_{\text{P}_2 \text{ choices}} \Big\} \text{P}_1 \text{ choices} \qquad B = \underbrace{\begin{bmatrix} -1 & 3 \\ 0 & -2 \end{bmatrix}}_{\text{P}_2 \text{ choices}} \Big\} \text{P}_1 \text{ choices}$$

To find a mixed Nash equilibrium, we need to compute vectors

$$y^* := [\, y_1^* \quad 1 - y_1^* \,]', \qquad y_1^* \in [0, 1]$$
$$z^* := [\, z_1^* \quad 1 - z_1^* \,]' \qquad z_1^* \in [0, 1],$$

for which

$$y^{*\prime} A z^* = y_1^*(1 - 6z_1^*) + 4z_1^* - 1 \leq y_1(1 - 6z_1^*) + 4z_1^* - 1, \quad \forall y_1 \in [0, 1] \quad (10.1)$$
$$y^{*\prime} B z^* = z_1^*(2 - 6y_1^*) + 5y_1^* - 2 \leq z_1(2 - 6y_1^*) + 5y_1^* - 2, \quad \forall z_1 \in [0, 1]. \quad (10.2)$$

Attention! This technique would have failed if (10.3) would have led to y_1^* or z_1^* not in the interval $[0, 1]$. In that case, we would have had to find a mixed Nash equilibrium for which (10.1)–(10.2) hold without constant right-hand sides.

This can be achieved if we are able to make the right hand side of (10.1) independent of y_1 and the right-hand side of (10.2) independent of z_1. In particular, by making

$$\begin{cases} 1 - 6z_1^* = 0 \\ 2 - 6y_1^* = 0 \end{cases} \Leftrightarrow \begin{cases} z_1^* = \frac{1}{6} \\ y_1^* = \frac{1}{3}. \end{cases} \quad (10.3)$$

This leads to the following mixed Nash equilibrium and outcomes

$$(y^*, z^*) = \underbrace{([\tfrac{1}{3} \ \ \tfrac{2}{3}]', [\tfrac{1}{6} \ \ \tfrac{5}{6}]')}_{\substack{\mathsf{P}_1 \text{ (husband) goes to football 66\% of times} \\ \mathsf{P}_2 \text{ (wife) goes to football 83\% of times}}} \tag{10.4a}$$

$$(y^{*\prime}Az^*, y^{*\prime}Bz^*) = (4z_1^* - 1, 5y_1^* - 2) = \left(-\tfrac{1}{3}, -\tfrac{1}{3}\right). \tag{10.4b}$$

Note. It turns out that (10.4), together with the pure equilibria (1, 1) and (2, 2), are the only Nash equilibria for this game. In this game, the mixed equilibrium (10.4) is the only one that is *not* admissible in the sense of Definition 9.3.

This particular Nash equilibrium has the very special property that

$$y^{*\prime}Az^* = y'Az^* \quad \forall y \in \mathcal{Y} \qquad y^{*\prime}Bz^* = y^{*\prime}Bz, \quad \forall z \in \mathcal{Z}$$

and therefore it is also a Nash equilibrium for a bimatrix game defined by the matrices $(-A, -B)$, which correspond to exactly opposite objectives by both players.

Note. However, for the game $(-A, -B)$ the pure Nash equilibria are different: (1, 2) with outcomes $(0, -3)$ and (2, 1) with outcomes $(-3, 0)$. In this case, the mixed Nash equilibrium has outcomes $(\frac{1}{3}, \frac{1}{3})$ and is also admissible.

This property of some Nash equilibria is not as unusual as one could imagine at first and, in fact, motivates the following definition:

Definition 10.1 (completely mixed Nash equilibria). A mixed Nash equilibrium (y^*, z^*) is said to be *completely mixed* or an *inner-point equilibria* if all probabilities are strictly positive, i.e., $y_i^*, z_j^* \in (0, 1), \forall i, j$.

Notation. For zero-sum games, i.e., when $A = -B$, we say that (y^*, z^*) is a *completely mixed saddle-point equilibrium.*

It turns out that all completely mixed Nash equilibria have the property we encountered before:

Lemma 10.1 (completely mixed Nash equilibria). If (y^*, z^*) is a completely mixed Nash equilibrium with outcomes (p^*, q^*) for a bimatrix game defined by the matrices (A, B), then

Notation. We denote by $\mathbf{1}_{k\times 1}$ a column vector with k entries, all equal to one.

$$Az^* = p^*\mathbf{1}_{m\times 1}, \qquad B'y^* = q^*\mathbf{1}_{n\times 1}. \tag{10.5}$$

Consequently, (y^*, z^*) is also a mixed Nash equilibrium for the three bimatrix games defined by $(-A, -B)$, $(A, -B)$, and $(-A, B)$. □

Note. From equations (10.5) we can also conclude that

$$y'Az^* = p^*, \quad \forall y$$
$$y^{*\prime}Bz = q^*, \quad \forall z,$$

which means that when P_1 uses their mixed Nash policy, the outcome for P_2 will not depend on P_2's policy; and vice versa.

Proof of Lemma 10.1. Assuming that (y^*, z^*) is a completely mixed Nash equilibrium for the game defined by (A, B), we have that

$$y^{*\prime}Az^* = \min_y y'Az^* = \min_y \sum_i y_i \underbrace{(Az^*)_i}_{i\text{th row of } Az^*}.$$

If one row i of Az^* was strictly larger than any of the remaining ones, then the minimum would be achieved with $y_i = 0$ and the Nash equilibrium would not be completely mixed. Therefore to have a completely mixed equilibrium, we need to have all the rows of Az^* exactly equal to each other:

$$Az^* = p^*\mathbf{1}_{m\times 1},$$

for some scalar p^*, which means that

$$y^{*\prime}Az^* = y'Az^* = p^*, \quad \forall y, y^* \in \mathcal{Y}. \tag{10.6}$$

Similarly, since none of the $z_j = 0$, we can also conclude that all columns of $y^{*\prime}B$ (which are the rows of $B'y^*$) must all be equal to some constant q^* and therefore

$$y^{*\prime}Bz^* = y^{*\prime}Bz = q^*, \quad \forall z, z^* \in \mathcal{Y}. \tag{10.7}$$

From (10.6)–(10.7), we conclude (p^*, q^*) is indeed the Nash outcome of the game and that (y^*, z^*) is also a mixed Nash equilibrium for the three bimatrix games defined by $(-A, -B)$, $(A, -B)$, and $(-A, B)$. ■

This result is also interesting for zero-sum games:

Example. The unique mixed saddle-point equilibrium of the Rock-Paper-Scissors game that we found in Example 4.2 is completely mixed.

Corollary 10.1. If (y^*, z^*) is a completely mixed saddle-point equilibrium for the zero-sum matrix game defined by A with mixed value p^*, then

$$Az^* = p^*\mathbf{1}_{m\times 1}, \qquad A'y^* = p^*\mathbf{1}_{n\times 1},$$

for some scalar p^*. Consequently, (y^*, z^*) is also a mixed saddle-point equilibrium for the zero-sum matrix game defined by $-A$, and a mixed Nash equilibrium for the two (non-zero-sum) bimatrix games defined by (A, A) and $(-A, -A)$. □

10.2 Computation of Completely Mixed Nash Equilibria

The computation of completely mixed Nash equilibria is particularly simple because, as we saw in Lemma 10.1, all these equilibria must satisfy

$$B'y^* = q^*\mathbf{1}_{n\times 1}, \qquad \mathbf{1}'y^* = 1, \qquad Az^* = p^*\mathbf{1}_{m\times 1}, \qquad \mathbf{1}'z^* = 1, \tag{10.8}$$

which provides a linear system with $n + m + 2$ equations and an equal number of unknowns: m entries of y^*, n entries of z^*, and the two scalars p^* and q^*.

After solving (10.8), we must still verify that the resulting y^* and z^* do have nonzero entries so that they belong to the sets $\mathcal{Y}$ and $\mathcal{Z}$, respectively. It turns out that if they do, we can immediately conclude that we have found a Nash equilibrium.

Note. The equilibrium (y^*, z^*) in Lemma 10.2 will be completely mixed if all entries of y^* and z^* are strictly positive, otherwise it is just a mixed Nash equilibrium.

Lemma 10.2. Suppose that the vectors (y^*, z^*) satisfy (10.8). If all entries of y^* and z^* are nonnegative, then (y^*, z^*) is a mixed Nash equilibrium. □

This result is actually a corollary of a more general result (Theorem 10.1) that we will state shortly, so we will not prove Lemma 10.2 at this point.

Note. See Exercise 10.2. ▷ p. 126

Attention! In view of Lemma 10.1, completely mixed Nash equilibria must satisfy (10.8) so the solutions to these equations actually provide us with the whole set of potential completely mixed Nash equilibria.

Example 10.1 (Battle of the sexes). For the battle of the sexes game introduced in Section 10.1, equation (10.8) becomes

$$Az^* = \begin{bmatrix} -2 & 0 \\ 3 & -1 \end{bmatrix} \begin{bmatrix} z_1^*, \\ 1 - z_1^* \end{bmatrix} = \begin{bmatrix} p^* \\ p^* \end{bmatrix},$$

$$B'y^* = \begin{bmatrix} -1 & 0 \\ 3 & -2 \end{bmatrix} \begin{bmatrix} y_1^* \\ 1 - y_1^* \end{bmatrix} = \begin{bmatrix} q^* \\ q^* \end{bmatrix},$$

which is equivalent to

$$p^* = -2z_1^*, \quad 4z_1^* = 1 + p^*, \quad q^* = -y_1^*, \quad 5y_1^* = 2 + q^*$$
$$\Rightarrow z_1^* = \frac{1}{6}, \quad y_1^* = \frac{1}{3}, \quad p^* = q^* = -\frac{1}{3},$$

as we had previously concluded. However, we now know that this is the unique completely mixed Nash equilibrium for this game.

If we consider, instead, the version of the battle of the sexes in Example 9.3, equation (10.8) now becomes

$$Az^* = \begin{bmatrix} -2 & 1 \\ 0 & -1 \end{bmatrix} \begin{bmatrix} z_1^*, \\ 1 - z_1^* \end{bmatrix} = \begin{bmatrix} p^* \\ p^* \end{bmatrix},$$
$$B'y^* = \begin{bmatrix} -1 & 3 \\ 2 & -2 \end{bmatrix} \begin{bmatrix} y_1^* \\ 1 - y_1^* \end{bmatrix} = \begin{bmatrix} q^* \\ q^* \end{bmatrix},$$

which is equivalent to

$$-3z_1^* + 1 = p^*, \quad z_1^* - 1 = p^*, \quad -4y_1^* + 3 = q^*, \quad 4y_1^* - 2 = q^*$$
$$\Rightarrow z_1^* = \frac{1}{2}, \quad y_1^* = \frac{5}{8}, \quad p^* = -\frac{1}{2}, \quad q^* = \frac{1}{2}.$$

Note. Recall that (1, 1) was a pure Nash equilibrium with outcomes $(-2, -1)$ and (2, 2) the other pure Nash equilibrium with outcomes $(-1, -2)$.

This shows that for the version of the game considered in Example 9.3, the only completely mixed Nash equilibrium is not admissible, as it is strictly worse for both players than the pure equilibrium that we found before. □

10.3 Numerical Computation of Mixed Nash Equilibria

We now consider a systematic numeric procedure to find mixed Nash equilibria for a bimatrix game defined by two $m \times n$ matrices $A = [a_{ij}]$ and $B = [b_{ij}]$ expressing the outcomes of players P_1 and P_2, respectively. As usual, P_1 selects a row of A/B and P_2 selects a column of A/B. Mixed Nash equilibria for this game can be found by solving a "quadratic program":

Theorem 10.1. The pair of policies $(y^*, z^*) \in \mathcal{Y} \times \mathcal{Z}$ is a mixed Nash equilibrium with outcome (p^*, q^*) if and only if the tuple (y^*, z^*, p^*, q^*) is a (global) solution to the following minimization:

Notation. An optimization such as (10.9)—in which one minimizes a quadratic function subject to affine equality and inequality constraints—is called a *quadratic program.*

MATLAB® Hint 3. Quadratic programs can be solved numerically with MATLAB® using `quadprog` from the Optimization toolbox. ▷ p. 122

$$\underbrace{\begin{aligned} &\text{minimize} && \text{y'(A+B)z-p-q} \\ &\text{subject to} && Az \succeq p\mathbf{1} \\ &&& B'y \succeq q\mathbf{1} \\ &&& \left.\begin{matrix} y \succeq 0 \\ \mathbf{1}y = 1 \end{matrix}\right\} \; (y \in \mathcal{Y}) \\ &&& \left.\begin{matrix} z \succeq 0 \\ \mathbf{1}z = 1 \end{matrix}\right\} \; (z \in \mathcal{Z}) \end{aligned}}_{\text{optimization over } m+n+2 \text{ parameters } (y_1, y_2, \dots, y_m, z_1, z_2, \dots, z_n, p, q)} \tag{10.9}$$

Moreover, this minimization always has a global minima at zero. □

For zero-sum games $A = -B$ and (10.9) is actually a linear program. However, this linear program is different than the one that we found in Lecture 6. This one finds both policies in one shot, but has more optimization variables and more constraints. It is generally less efficient because solving two small problems is better than solving a large one.

Notation. A quadratic form is called *indefinite* if it can take both positive and negative values.

Attention! Unless $A = -B$, which would correspond to a zero-sum game, the quadratic criteria in (10.9) is indefinite because we can select z to be any vector for which $(A + B)z \neq 0$ and then obtain a positive value with

$$y = (A + B)z, \;\; p = q = 0 \Rightarrow y'(A + B)'(A + B)z - p - q = \;\; \|(A + B)z\|^2 > 0$$

and a negative value with

$$y = -(A + B)z, \;\; p = q = 0$$
$$\Rightarrow y'(A + B)'(A + B)z - p - q = -\|(A + B)z\|^2 < 0.$$

Note. We will encounter in Lecture 13 an iterative technique called *fictitious play* that can also be used to compute mixed Nash equilibrium for many bimatrix games. ▷ p. 156

In this case, the quadratic criteria in (10.9) is not convex, which means that numerical solvers can easily get caught in local minima. It is therefore important to verify that the solver found a global minimum. This can be easily done since we know that the global minimum is exactly equal to zero. When a solver gets caught in a local minimum, one may need to restart it at a different (typically random) initial point.

MATLAB® Hint 3 (Quadratic programs). The command

```
[x,val]=quadprog(H,c,Ain,bin,Aeq,beq,low,high,x0)
```

from MATLAB®'s Optimization Toolbox numerically solves quadratic programs of the form

Attention! When H is not positive semidefinite, the quadratic program is not convex. This is always the case for mixed Nash equilibria unless the game is zero sum, i.e., unless $A = -B$.

$$\begin{aligned} \text{minimum} \quad & \tfrac{1}{2}x'\mathtt{H}x + \mathtt{c}'x \\ \text{subject to} \quad & \mathtt{Ainx} \dot{\leq} \mathtt{bin} \\ & \mathtt{Aeqx} = \mathtt{beq} \\ & \mathtt{low} \dot{\leq} x \dot{\leq} \mathtt{high} \end{aligned}$$

and returns the value `val` of the minimum and a vector `x` that achieves the minimum.

The vector `low` can have some or all entries equal to `-Inf` and the vector `high` have some or all entries equal to `+Inf` to avoid the corresponding inequality constraints.

The (optional) vector `x0` provides a starting point for the numerical optimization. This is particularly important when `H` is indefinite since in this case the minimization is not convex and may have local minima.

The MATLAB® command `optimset` can be used to set optimization options for `quadprog`. In particular, it allows the selection of the optimization algorithm, which for the computation of Nash equilibria, must support non-convex problems.

The following MATLAB® code can be used to find a mixed Nash equilibrium to the bimatrix game defined by A and B, starting from a random initial condition x0.

Note. The quadratic criteria in (10.9) can be written as $\frac{1}{2}\begin{bmatrix} y' & z' \end{bmatrix} \begin{bmatrix} 0 & A+B \\ A'+B' & 0 \end{bmatrix} \begin{bmatrix} y \\ z \end{bmatrix} - p - q.$

Note. This code found the completely mixed (but not admissible) Nash equilibrium for the battle of the sexes game with
A=[-2,0;3,-1];
B=[-1,3;0,-2];
For the prisoners' dilemma game with
A=[2,30;0,8];
B=[2,0;30,8];
it very often returns a local minimum that is *not a mixed Nash equilibrium*. When this happens the code must be re-run with a different initial point x0.

```
[m,n]=size(A);
x0=rand(n+m+2,1);
% y'(A+b)z-p-q
H=[zeros(m,m),A+B,zeros(m,2);A'+B',zeros(n,n+2);zeros(2,m+n+2)];
c=[zeros(m+n,1);-1;-1];
% A z>=p & B' y >= q
Ain=[zeros(m,m),-A,ones(m,1),zeros(m,1);-B',zeros(n,n+1),ones(n,1)];
bin=zeros(m+n,1);
% sum(y)=sum(z)=1
Aeq=[ones(1,m),zeros(1,n+2);zeros(1,m),ones(1,n),0,0];
beq=[1;1];
% y_i, z_i in [0,1]
low=[zeros(n+m,1);-inf;-inf];
high=[ones(n+m,1);+inf;+inf];
% solve quadratic program
options=optimset('TolFun',1e-8,'TolX',1e-8,'TolPCG',1e-8,...
                 'Algorithm','active-set');
[x,val,exitflag]=quadprog(H,c,Ain,bin,Aeq,beq,low,high,x0,options)
y=x(1:m)
z=x(m+1:m+n)
p=x(m+n+1)
q=x(m+n+2)
```

□

Proof of Theorem 10.1. This theorem makes two statements that need to be justified separately: a mixed Nash equilibrium is always a global minimum to (10.9) and also that any global minimum to (10.9) is a mixed Nash equilibrium.

We start by proving the first statement so we assume that (y^*, z^*) is a mixed Nash equilibrium with outcome equal to (p^*, q^*) and we will show that (y^*, z^*, p^*, q^*) is a global minimum to (10.9). To this effect we need to show the following:

1. The point (y^*, z^*, p^*, q^*) satisfies all the constraints in (10.9). Indeed, since

$$p^* = y^{*\prime} A z^* \leq y' A z^*, \quad \forall y \in \mathcal{Y},$$

in particular for every integer $i \in \{1, 2, \ldots, m\}$ if we pick

$$y = \underbrace{\begin{bmatrix} 0 & \cdots & 0 & 1 & 0 & \cdots & 0 \end{bmatrix}'}_{1 \text{ at the } i\text{th position}}$$

we conclude that

$$p^* \leq \underbrace{(Az^*)_i}_{i\text{th entry of } Az^*}$$

and therefore $p^*\mathbf{1} \stackrel{.}{\leq} Az^*$. On the other hand, since

$$q^* = y^{*\prime} B z^* \leq y^{*\prime} B z, \quad \forall z \in \mathcal{Z},$$

we also conclude that $q^*\mathbf{1}' \lneq y^{*\prime}B$. This entry-wise inequality between row vectors can be converted into an entry-wise inequality between column vectors by transposition: $q^*\mathbf{1} \lneq B'y^*$. The remaining constraints on y^* and z^* that appear in (10.9) hold because $y^* \in \mathcal{Y}$ and $z^* \in \mathcal{Z}$.

2. The (y^*, z^*, p^*, q^*) achieves the global minimum, which is equal to zero. To show this, we first note that since $p^* = y^{*\prime}Az^*$ and $q^* = y^{*\prime}Bz^*$ we indeed have that

$$y^{*\prime}(A+B)z^* - p^* - q^* = 0.$$

It remains to show that no other vectors y and z that satisfy the constraints can lead to a value for the criteria lower than zero:

$$\begin{cases} Az \geq p\mathbf{1} \\ B'y \geq q\mathbf{2} \end{cases} \Rightarrow \begin{cases} y'Az \geq p \\ z'B'y \geq q \end{cases} \Rightarrow y'(A+B)z - p - q \geq 0.$$

To prove the converse statement we assume that (y^*, z^*, p^*, q^*) is a global minimum to (10.9) and we will show that (y^*, z^*) must be a mixed Nash equilibrium with outcome (p^*, q^*).

From the basic existence Theorem 9.1 we know that there is at least one mixed Nash equilibrium and we have seen above that this leads to a global minimum of (10.9) equal to zero. Therefore any global minimum (y^*, z^*, p^*, q^*) to (10.9) must satisfy:

$$y^{*\prime}(A+B)z^* - p^* - q^* = 0 \tag{10.10}$$

$$Az^* \gtrdot p^*\mathbf{1}, \quad z^* \in \mathcal{Z} \tag{10.11}$$

$$B'y^* \gtrdot q^*\mathbf{1}, \quad y^* \in \mathcal{Y}. \tag{10.12}$$

From (10.11)–(10.12) we conclude that

$$y'Az^* \geq p^*, \quad \forall y \in \mathcal{Y}, \qquad z'B'y^* \geq q^*, \quad \forall z \in \mathcal{Z}. \tag{10.13}$$

The proof is completed as soon as we show that

$$y^*Az^* = p^*, \qquad z^{*\prime}B'y^* = q^*. \tag{10.14}$$

To achieve this, we set $y = y^*$ and $z = z^*$ in (10.13), which leads to

$$y^*Az^* - p^* \geq 0, \qquad z^{*\prime}B'y^* - q^* \geq 0.$$

However, because of (10.10) the two numbers in the left-hand sides of these inequalities must add up to zero so they must both necessarily be equal to zero, which confirms that (10.14) holds and completes the proof. ■

10.4 Practice Exercise

10.1. Use MATLAB® to compute Nash equilibrium policies for the following two games:

1. battle of the sexes with

$$A = \underbrace{\begin{bmatrix} -2 & 0 \\ 3 & -1 \end{bmatrix}}_{\text{P}_2\text{ choices}} \Big\} \text{P}_1 \text{ choices} \qquad B = \underbrace{\begin{bmatrix} -1 & 3 \\ 0 & -2 \end{bmatrix}}_{\text{P}_2\text{ choices}} \Big\} \text{P}_1 \text{ choices}$$

2. prisoners' dilemma

$$A = \underbrace{\begin{bmatrix} 2 & 30 \\ 0 & 8 \end{bmatrix}}_{\text{P}_2\text{ choices}} \Big\} \text{P}_1 \text{ choices} \qquad B = \underbrace{\begin{bmatrix} 2 & 0 \\ 30 & 8 \end{bmatrix}}_{\text{P}_2\text{ choices}} \Big\} \text{P}_1 \text{ choices.}$$

Solution to Exercise 10.1. The following MATLAB® code can be used to solve both problems.

```
[m,n]=size(A);
x0=rand(n+m+2,1);
% y'(A+b)z-p-q
H=[zeros(m,m),A+B,zeros(m,2);
   A'+B',zeros(n,n+2);zeros(2,m+n+2)];
c=[zeros(m+n,1);-1;-1];
% A z>=p & B' y >= q
Ain=[zeros(m,m),-A,ones(m,1),zeros(m,1);
     -B',zeros(n,n+1),ones(n,1)];
bin=zeros(m+n,1);
% sum(y)=sum(z)=1
Aeq=[ones(1,m),zeros(1,n+2);zeros(1,m),ones(1,n),0,0];
beq=[1;1];
% y_i, z_i in [0,1]
low=[zeros(n+m,1);-inf;-inf];
high=[ones(n+m,1);+inf;+inf];
% solve quadratic program
options=optimset('TolFun',1e-8,'TolX',1e-8,'TolPCG',1e-8,...
                 'Algorithm','active-set');
[x,val,exitflag]=quadprog(H,c,Ain,bin,...
                          Aeq,beq,low,high,x0,options)
y=x(1:m)
z=x(m+1:m+n)
p=x(m+n+1)
q=x(m+n+2)
```

1. For the battle of the sexes game, the code above should be preceded by

```
A=[-2,0;3,-1];
B=[-1,3;0,-2];
```

and we obtain the following global minimum

```
y = 0.3333
    0.6667
z = 0.1667
    0.8333
p = -0.3333
q = -0.3333
```

2. For the prisoners' dilemma game the code above should be preceded by

   ```
   A=[2,30;0,8];
   B=[2,0;30,8];
   ```

   and we obtain the following global minimum

   ```
   y = 0.0000
       1.0000
   z = 0.0000
       1.0000
   p = 8.0000
   q = 8.0000
   ```

   Only about 15% of the time that we ran the above optimization did we get a global minima. In the remaining 85% of the cases, we obtained local minima.

10.5 Additional Exercise

10.2 (Completely mixed Nash equilibria). Prove Lemma 10.2 using Theorem 10.1.

LECTURE 11

N-Player Games

In this lecture we extend to games with N-players several of the concepts introduced for two-player games.

11.1 N-PLAYER GAMES

The notions introduced for bimatrix games can be extended to games with any number of players. In this lecture we consider general games with N *players* $\mathsf{P}_1, \mathsf{P}_2, \ldots, \mathsf{P}_N$, which are allowed to select policies within *action spaces* $\Gamma_1, \Gamma_2, \ldots, \Gamma_N$. When

$$\begin{cases} \mathsf{P}_1 \text{ uses policy } \gamma_1 \in \Gamma_1 \\ \mathsf{P}_2 \text{ uses policy } \gamma_2 \in \Gamma_2 \\ \vdots \\ \mathsf{P}_N \text{ uses policy } \gamma_N \in \Gamma_N, \end{cases}$$

we denote by

$$J_i(\gamma_1, \gamma_2, \ldots, \gamma_N) \tag{11.1}$$

the *outcome of the game* for player P_i. Each player wants to *minimize* their own outcome, and does not care about the outcome of the other players.

Notation 3. To avoid writing a long series of policies it is common to use γ_{-i} to denote all but the ith policy and rewrite (11.1) as $J_i(\gamma_i, \gamma_{-i})$. ▷ p. 127

Notation 3. To avoid writing a long series of policies as in the argument of J_i in (11.1), it is common to use certain abbreviations when writing equations for N-player games. An obvious abbreviation consists of denoting the list of all policies simply by

$$\gamma \equiv (\gamma_1, \gamma_2, \ldots, \gamma_N),$$

and the Cartesian product of all action spaces simply by

$$\Gamma \equiv \Gamma_1 \times \Gamma_2 \times \cdots \times \Gamma_N.$$

The following less intuitive but extremely useful abbreviation is used to denote a list of all but the ith policy:

$$\gamma_{-i} \equiv (\gamma_1, \gamma_2, \ldots, \gamma_{i-1}, \gamma_{i+1}, \ldots, \gamma_N).$$

Consistent with this terminology, it is often convenient to separate the dependence of J_i on γ_i and on the remaining policies γ_{-i} and write (11.1) as follows:

$$J_i(\gamma_i, \gamma_{-i}). \tag{11.2}$$

This notation can lead to some ambiguity, since (11.2) is suppose to represent exactly the same as (11.1) and does *not* imply any change in the order of the arguments. In particular (11.2) represents (11.1) and *not*

$$J_i(\gamma_i, \gamma_1, \gamma_2, \ldots, \gamma_{i-1}, \gamma_{i+1}, \ldots, \gamma_N),$$

which would imply that P_1 uses the policy γ_i, P_2 uses the policy γ_1, and so on.

This terminology also applies to action spaces, as in

$$\gamma_{-i} \in \Gamma_{-i}$$

which is meant to be a short-hand notation for

$$\gamma_1 \in \Gamma_1, \gamma_2 \in \Gamma_2, \ldots, \gamma_{i-1} \in \Gamma_{i-1}, \gamma_{i+1} \in \Gamma_{i+1}, \ldots, \gamma_N \in \Gamma_N. \quad \square$$

11.1.1 Security Levels and Policies

The definition of security policies for N-player games captures the notion of finding the policy that guarantees the least possible cost, assuming the worse possible choice (rational or not) by the other players.

Notation. The *infimum* of a set is its largest lower bound and the *supremum* of a set is its smallest upper bound. ▷ p. 44

Notation. The security level for P_i only depends on J_i, this is emphasized by only using J_i as the argument for $\bar{V}_{\Gamma_1,\Gamma_2}(\cdot)$.

Definition 11.1 (Security policy). The *security level* for P_i, $i \in \{1, 2, \ldots, N\}$ is defined by

$$\bar{V}(J_i) := \underbrace{\inf_{\gamma_i \in \Gamma_i}}_{\substack{\text{minimize cost assuming} \\ \text{worst choice by } \mathsf{P}_i}} \underbrace{\sup_{\gamma_{-i} \in \Gamma_{-i}}}_{\substack{\text{worst choice by all other players } \mathsf{P}_{-j}, \\ \text{from } \mathsf{P}_i\text{'s perspective}}} J_i(\gamma_i, \gamma_{-i})$$

Note. In general, security policies may not exist because the infimum may not be achieved by a policy in Γ_i.

Notation. Equation (11.3) is often written as $\gamma_i^* \in \arg\min_{\gamma_i} \sup_{\gamma_{-i}} J_i(\gamma_i, \gamma_{-i})$. The use of "$\in$" instead of "$=$" emphasizes that there may be several (or none) γ_i^* that achieve the infimum.

and a *security policy* for P_i is any policy $\gamma_i^* \in \Gamma_i$ for which the infimum above is achieved, i.e.,

$$\bar{V}(J_i) := \inf_{\gamma_i \in \Gamma_i} \sup_{\gamma_{-i} \in \Gamma_{-i}} J_i(\gamma_i, \gamma_{-i}). = \underbrace{\sup_{\gamma_{-i} \in \Gamma_{-i}} J_i(\gamma_i^*, \gamma_{-i}^*)}_{\gamma_i^* \text{ achieves the infimum}}. \tag{11.3}$$

An N-tuple of policies $(\gamma_1^*, \gamma_2^*, \ldots, \gamma_N^*)$ is said to be *minimax* if each γ_i is a security policy for P_i.

11.1.2 Nash Equilibria

The definition of equilibria for non-zero-sum games captures the notion of no regret in the sense that after knowing the choice made by the other player, each player finds that their own policy provided the lowest possible cost against the choice of the other player.

Definition 11.2 (Nash equilibrium). An N-tuple of policies $\gamma^* := (\gamma_1^*, \gamma_2^*, \ldots, \gamma_N^*) \in \Gamma_1 \times \Gamma_2 \times \cdots \times \Gamma_N$ is called a *Nash equilibrium* if

Note. These conditions are consistent with the Meta Definition 1.1.

$$J_i(\gamma^*) = J_i(\gamma_i^*, \gamma_{-i}^*) \leq J_i(\gamma_i, \gamma_{-i}^*), \quad \forall \gamma_i \in \Gamma_i, \ \ i \in \{1, 2, \ldots, N\}$$

and the N-tuple $(J_1(\gamma^*), J_2(\gamma^*), \ldots, J_N(\gamma^*))$ is called the *Nash outcome of the game*. The Nash equilibrium is said to be *admissible* if there is no "better" Nash equilibrium in the sense that there is no other Nash equilibrium $\bar{\gamma}^* := (\bar{\gamma}_1^*, \bar{\gamma}_2^*, \ldots, \bar{\gamma}_N^*) \in \Gamma_1 \times \Gamma_2 \times \cdots \times \Gamma_N$ such that

Note. All players do no worse with $\bar{\gamma}^*$ and at least one does strictly better.

$$J_i(\bar{\gamma}^*) \leq J_i(\gamma^*), \quad \forall i \in \{1, 2, \ldots, N\}$$

with a strict inequality for at least one player.

11.2 Pure N-Player Games in Normal Form

Pure n-player games in normal form are played by N players $\mathsf{P}_1, \mathsf{P}_2, \ldots, \mathsf{P}_N$, each selecting policies from finite *action spaces*:

$$\mathsf{P}_i \text{ has available } m_i \text{ actions: } \Gamma_i := \{1, 2, \ldots, m_i\}.$$

The *outcomes* for the players are quantified by N tensors $A^1, A^2, \ldots, A^N$, each N-dimensional with dimensions $(m_1, m_2, \ldots, m_N)$. When

Notation. In mathematics, a *tensor* is essentially a multi-dimensional array that generalizes the concept of matrix for dimensions higher than two.

$$\begin{cases} \mathsf{P}_1 \text{ selects action } k_1 \in \Gamma_1 := \{1, 2, \ldots, m_1\} \\ \mathsf{P}_2 \text{ selects action } k_2 \in \Gamma_2 := \{1, 2, \ldots, m_2\} \\ \vdots \\ \mathsf{P}_N \text{ selects action } k_N \in \Gamma_N := \{1, 2, \ldots, m_N\}, \end{cases}$$

the *outcome* for the ith player is obtained from the appropriate entry $a^i_{k_1 k_2 \cdots k_N}$ of the tensor A^i. In this formulation, the understanding is that all players want to *minimize* their respective outcomes.

Notation. To distinguish the different tensors for each player P_i we use a super-script ^{i}. This leads to some ambiguity since a super-script could be confused with a power, however this is needed since sub-scripts will be needed to index the entries of the tensor.

Testing if a particular N-tuple of pure policies $(k_1^*, k_2^*, \ldots, k_N^*)$ is a Nash equilibrium is straightforward as one simply needs to check if

$$a^i_{k_i^* \, k_{-i}^*} \leq a^i_{k^i \, k_{-i}^*}, \quad \forall k^i \in \{1, 2, \ldots, m_i\}, \ \ \forall i \in \{1, 2, \ldots, N\}.$$

However, finding a Nash equilibrium in pure policies is generally computationally difficult, as one potentially needs to check all possible N-tuples, which are as many as

$$m_1 \times m_2 \times \cdots \times m_N.$$

Note. As in zero-sum or bimatrix games, there may be no pure Nash equilibria for an N-player game in normal form.

11.3 Mixed Policies for N-Player Games in Normal Form

N-player games also admit *mixed policies*. As before, the idea behind mixed policies is that the players select their actions randomly according to previously selected probability distributions. A *mixed policy for player* P_i is thus a set of numbers

$$y^i := (y^i_1, y^i_2, \ldots, y^i_{m_i}), \quad \sum_{k=1}^{m_i} y^i_k = 1 \quad y^i_k \geq 0, \quad \forall k \in \{1, 2, \ldots, m_i\},$$

with the understanding that y^i_k is the probability that P_i uses to select the action $k \in \{1, 2, \ldots, m_i\}$. Each mixed policy y^i is thus an element of the action space $\mathcal{Y}^i$ consisting of probability distributions over m_i actions.

It is assumed that the random selections by all players are done statistically independently and each player P_i tries to minimize their own *expected outcome*, which is given by

$$J_i = \sum_{k_1=1}^{m_1} \sum_{k_2=1}^{m_2} \cdots \sum_{k_N=1}^{m_N} \underbrace{y^1_{k_1} y^2_{k_2} \cdots y^N_{k_N}}_{\substack{\text{probability that } \mathsf{P}_1 \text{ selects } k_1 \\ \text{and } \mathsf{P}_2 \text{ selects } k_2 \text{ and } \ldots}} \quad \underbrace{a^i_{k_1 k_2 \cdots k_N}}_{\substack{\text{outcome when } \mathsf{P}_1 \text{ selects } k_1 \\ \text{and } \mathsf{P}_2 \text{ selects } k_2 \text{ and } \ldots}} \tag{11.4}$$

Notation. It is often convenient to write (11.4) in the following compressed form:

$$\sum_{k_1 k_2 \cdots k_N} \left(\prod_j y^j_{k_j}\right) a^i_{k_1 k_2 \cdots k_N}.$$

Having defined mixed policies for an N-player game in normal form, we can use the concepts of security levels, security policies, and saddle-point equilibria introduced in Section 11.1 for general games. In particular, this leads to the following definition of a saddle-point equilibrium:

Note. One can also define average security levels and mixed policies, which never lead to worse outcomes and often result in strictly lower costs (cf. Exercise 9.3). This means that players can always protect themselves better (or at least no worse) with mixed policies (even when the other players are also allowed to use mixed policies).

Definition 11.3 (Mixed Nash equilibrium). An N-tuple of policies $(y^{1*}, y^{2*}, \ldots, y^{N*}) \in calligY^1 \times \mathcal{Y}^2 \times \cdots \times \mathcal{Y}^N$ is called a *mixed Nash equilibrium* if

$$\sum_{k_1} \sum_{k_2} \cdots \sum_{k_N} \boxed{y^{1*}_{k_1}} y^{2*}_{k_2} \cdots y^{N*}_{k_N} a^1_{k_1 k_2 \cdots k_N}$$
$$\leq \sum_{k_1} \sum_{k_2} \cdots \sum_{k_N} \boxed{y^1_{k_1}} y^{2*}_{k_2} \cdots y^{N*}_{k_N} a^1_{k_1 k_2 \cdots k_N},$$
$$\sum_{k_1} \sum_{k_2} \cdots \sum_{k_N} y^{1*}_{k_1} \boxed{y^{2*}_{k_2}} y^{3*}_{k_3} \cdots y^{N*}_{k_N} a^2_{k_1 k_2 \cdots k_n}$$
$$\leq \sum_{k_1} \sum_{k_2} \cdots \sum_{k_N} y^{1*}_{k_1} \boxed{y^2_{k_2}} y^{3*}_{k_3} \cdots y^{N*}_{k_N} a^2_{k_1 k_2 \cdots k_n}, \ldots$$

or equivalently in a more compressed form

$$\sum_{k_1 k_2 \cdots k_N} \boxed{y^{i*}_{k_i}} \left(\prod_{j \neq i} y^{j*}_{k_j}\right) a^i_{k_1 k_2 \cdots k_N}$$
$$\leq \sum_{k_1 k_2 \cdots k_N} \boxed{y^i_{k_i}} \left(\prod_{j \neq i} y^{j*}_{k_j}\right) a^i_{k_1 k_2 \cdots k_N}, \quad \forall i \in \{1, 2, \ldots, N\}. \quad \square$$

As in bimatrix games, the introduction of mixed policies enlarges the action spaces for both players to the point that Nash equilibria now always exist:

Note. A proof of this result, due to Nash [10], can be constructed from Brouwer's Fixed-point Theorem.

Theorem 11.1 (Nash). Every N-player game in normal form has at least one mixed Nash equilibrium. □

11.4 Completely Mixed Policies

In general, computing Nash equilibria for N-player games in normal form is not an easy task. However, it can be simpler for games that admit completely mixed equilibria:

Definition 11.4 (completely mixed Nash equilibria). A mixed Nash equilibrium $(y^{1*}, y^{2*}, \dots, y^{N*})$ is said to be *completely mixed* or an *inner-point equilibrium* if all probabilities are strictly positive, i.e.,

$$y^{1*} \gtrdot 0,\ y^{2*} \gtrdot 0, \cdots, y^{N*} \gtrdot 0.$$

It turns out that all completely mixed Nash equilibria can be found by solving an algebraic multi-linear system of equations:

Notation. In (11.5), we are using k_{-i} to denote a summation over all indexes $k_1 k_2 \cdots k_{i-1} k_{i+1} \cdots k_N$.

Note. For games with more than two players (11.5) is no longer a linear system of equations and therefore solving it may not be easy. We will encounter in Lecture 13 an alternative technique called *fictitious play* that can be used to compute mixed Nash equilibria for many N-player games. ▹ p. 156

Lemma 11.1 (completely mixed Nash equilibri). If $(y^{1*}, y^{2*}, \dots, y^{N*})$ is a completely mixed Nash equilibrium with outcomes $(p^{1*}, p^{2*}, \dots, p^{N*})$ then

$$\sum_{k_{-i}} \left(\prod_{j \neq i} y^{j*}_{k_j} \right) a^i_{k_1 k_2 \cdots k_N} = p^{i*}, \quad \forall i \in \{1, 2, \dots, N\}. \tag{11.5}$$

Conversely, any solution $(y^{1*}, y^{2*}, \dots, y^{N*})$, $(p^1*, p^{2*}, \dots, p^{N*})$ to (11.5) for which

$$\sum_{k_i=1}^{m_i} y^{i*}_{k_i} = 1, \quad y^{i*} \gtrdot 0, \quad \forall i \in \{1, 2, \dots, N\} \tag{11.6}$$

corresponds to a mixed Nash equilibrium $(y^{1*}, y^{2*}, \dots, y^{N*})$ with outcomes $(p^{1*}, p^{2*}, \dots, p^{N*})$ for the original game, as well as for any similar game in which some (or all) players want to maximize instead of minimize their outcomes. □

Proof of Lemma 11.1. Assuming that $(y^{1*}, y^{2*}, \dots, y^{N*})$ is a completely mixed Nash equilibrium, we have that

$$\begin{aligned}\sum_{k_1 k_2 \cdots k_N} y^{i*}_{k_i} \left(\prod_{j \neq i} y^{j*}_{k_j} \right) a^i_{k_1 k_2 \cdots k_N} &= \min_{y^i} \sum_{k_1 k_2 \cdots k_N} y^i_{k_i} \left(\prod_{j \neq i} y^{j*}_{k_j} \right) a^i_{k_1 k_2 \cdots k_N} \\ &= \min_{y^i} \sum_{k_i} y^i_{k_i} \sum_{k_{-i}} \left(\prod_{j \neq i} y^{j*}_{k_j} \right) a^i_{k_1 k_2 \cdots k_N}.\end{aligned}$$

If one of the $\sum_{k_{-i}} \left(\prod_{j\neq i} y^{j*}_{k_j}\right) a^i_{k_1k_2\cdots k_N}$ was strictly larger than any of the remaining ones, then the minimum would be achieved with $y_i = 0$ and the Nash equilibrium would not be completely mixed. Therefore to have a completely mixed equilibrium, we necessarily must have (11.5) for some scalar p^{i*}.

Conversely, if $(y^{1*}, y^{2*}, \dots, y^{N*})$ and $(p^1*, p^{2*}, \dots, p^{N*})$ satisfy (11.5) and (11.6), then

$$\sum_{k_1 k_2 \cdots k_N} y^{i*}_{k_i} \left(\prod_{j\neq i} y^{j*}_{k_j}\right) a^i_{k_1k_2\cdots k_N} = \sum_{k_1 k_2 \cdots k_N} y^i{}_{k_i} \left(\prod_{j\neq i} y^{j*}_{k_j}\right) a^i_{k_1k_2\cdots k_N}$$
$$= \min_{y^i} \sum_{k_i} y^i_{k_i} p^{i*} = p^{i*}, \quad \forall y^i \in \mathcal{Y}^i,$$

which shows that $(y^{1*}, y^{2*}, \dots, y^{N*})$ is a mixed Nash equilibrium with outcome $(p^1*, p^{2*}, \dots, p^{N*})$. In fact, $(y^{1*}, y^{2*}, \dots, y^{N*})$ is also a mixed Nash equilibrium for a different game in which some (or all) players want to maximize instead of minimize the outcome. ■

LECTURE 12

Potential Games

In this lecture we introduce a special classes of N-player games called potential games. The terminology "potential games" was introduced by Monderer and Shapley [9], but the idea of characterizing a multiplayer game using a single "potential" function dates back from the work of Rosenthal [13] on congestion games. Nash equilibria are guaranteed to exist and are generally easy to find for these games.

12.1 Identical Interests Games

Consider a game with N players $\mathsf{P}_1, \mathsf{P}_2, \ldots, \mathsf{P}_N$, who select policies from the *action spaces* $\Gamma_1, \Gamma_2, \ldots, \Gamma_N$, respectively. As usual, when each P_i uses a policy $\gamma_i \in \Gamma_i$, we denote by

$$J_i(\gamma), \quad \gamma := (\gamma_1, \gamma_2, \ldots, \gamma_N) \in \Gamma := \Gamma_1 \times \Gamma_2 \times \cdots \Gamma_N$$

the *outcome of the game* for the player P_i and all players wanting to minimize their own outcomes.

A game is said to be of *identical interests* if all players have the same outcome, i.e., if there exists a function $\phi(\gamma)$ such that

$$J_i(\gamma) = \phi(\gamma), \quad \forall \gamma \in \Gamma, \;\; i \in \{1, 2, \ldots, N\}. \tag{12.1}$$

Note. The notion of a *directionally-local minimum* should not be confused with that of a *local minimum*, which would ask for the existence of a (possibly small) open set $\mathcal{O} \subset \Gamma$ containing γ^* for which $\phi(\gamma^*) \le \phi(\gamma), \forall \gamma \in \mathcal{O}$.

Note. Every global minimum is also a directionally-local minimum.

The notion of a Nash equilibrium for such games is closely related to the notion of minimum: A given $\gamma^* := (\gamma_1^*, \gamma_2^*, \ldots, \gamma_N^*) \in \Gamma$ is said to be a *global minimum of* ϕ if

$$\phi(\gamma^*) \le \phi(\gamma), \quad \forall \gamma \in \Gamma$$

and it is said to be a *directionally-local minimum of* ϕ if

$$\phi(\gamma_i^*, \gamma_{-i}^*) \le \phi(\gamma_i, \gamma_{-i}^*), \quad \forall \gamma_i \in \Gamma_i, \;\; i \in \{1, 2, \ldots, N\}. \tag{12.2}$$

Proposition 12.1. Consider an identical interests game with outcomes given by (12.1):

1. An N-tuple of policies $\gamma^* := (\gamma_1^*, \gamma_2^*, \ldots, \gamma_N^*) \in \Gamma$ is a Nash equilibrium if and only if γ^* is a directionally-local minimum of ϕ.
2. Assume that the potential function ϕ has at least one global minimum. An N-tuple of policies $\gamma^* := (\gamma_1^*, \gamma_2^*, \ldots, \gamma_N^*) \in \Gamma$ is an admissible Nash equilibrium if and only if γ^* is a global minimum to the function ϕ in (12.1). Moreover, all admissible Nash equilibria have the same value for all players, which is precisely the global minimum of ϕ. □

Attention! When the action spaces are finite, a global minimum always exists. In this case, identical interests games always have at least one admissible Nash equilibrium.

Proof of Proposition 12.1. Statement 1 is a direct consequence of the fact that for an identical interests game with outcomes given by (12.1) the definition of a Nash equilibrium is precisely the condition in (12.2).

To prove statement 2, suppose first that $\gamma^* := (\gamma_1^*, \gamma_2^*, \ldots, \gamma_N^*) \in \Gamma$ is a global minimum, then

$$\begin{aligned} J_i(\gamma_i^*, \gamma_{-i}^*) &:= \phi(\gamma_i^*, \gamma_{-i}^*) \leq \phi(\gamma_i, \gamma_{-i}^*) \\ &=: J_i(\gamma_i, \gamma_{-i}^*), \quad \forall \gamma_i \in \Gamma_i, \quad i \in \{1, 2, \ldots, N\}, \end{aligned}$$

which shows that γ^* is a Nash equilibrium. Moreover, since γ^* is a global minimum of ϕ, it must be admissible since no other policies could lead to a smaller value of the outcomes.

Conversely, suppose that $\gamma^* := (\gamma_1^*, \gamma_2^*, \ldots, \gamma_N^*) \in \Gamma$ is not a global minimum and that, instead, $\bar{\gamma}^* \neq \gamma^*$ is a global minimum. Then, from the reason above, we conclude that $\bar{\gamma}^*$ must be a Nash equilibrium that leads to outcomes that are "better" than those of γ^* for all players. This means that, even if γ^* is a Nash equilibrium, it cannot be admissible. ■

Attention! Identical interests games are especially attractive because all global minima of the common outcome ϕ of all players—often called the *social optima*—are admissible Nash equilibria. However, one needs to keep in mind the following two facts:

1. Even though all admissible Nash equilibria are global minima, there may be (non-admissible) Nash equilibria that are *not global minima.* So if the players choose non-admissible Nash equilibrium policies, they may still find themselves playing at an equilibrium that is not a global minimum.

   This may happen, e.g., with the pure bimatrix game defined by the matrices

   $$A = B = \begin{bmatrix} 0 & 2 \\ 2 & 1 \end{bmatrix},$$

   which has a single pure admissible Nash equilibrium (1, 1) that corresponds to the global minimum 0, but another non-admissible Nash equilibrium (2, 2) that corresponds to the outcome 1 for both players.
2. There may also be problems if multiple global minima exist, because even though all admissible Nash equilibria have the same outcome for all players, they *may not enjoy the order interchangeability property.*

This may happen, e.g., with the pure bimatrix game defined by the matrices

$$A = B = \begin{bmatrix} 0 & 1 \\ 1 & 0 \end{bmatrix},$$

which has two pure admissible Nash equilibria (1, 1) and (2, 2), both corresponding to the global minimum 0, but (1, 2) and (2, 1) are not global minima. □

12.2 Potential Games

Note. In words: in an exact potential game, this means that if a player P_i unilateral deviates from γ_i to $\bar{\gamma}_i$, then the change in their outcome is exactly equal to the change in the potential, which is common to all players.

The match between Nash equilibria and social optima holds for a class of games more general than identical interests games, which we shall explore next.

A game is said to be an *(exact) potential game* if there exists a function $\phi(\gamma_1, \gamma_2, \ldots, \gamma_N)$ such that

$$J_i(\gamma_i, \gamma_{-i}) - J_i(\bar{\gamma}_i, \gamma_{-i}) = \phi(\gamma_i, \gamma_{-i}) - \phi(\bar{\gamma}_i, \gamma_{-i}),$$
$$\forall \gamma_i, \bar{\gamma}_i \in \Gamma_i, \quad \gamma_{-i} \in \Gamma_{-i}, \quad i \in \{1, 2, \ldots, N\} \tag{12.3}$$

Note. In words: in an ordinal potential game, if a player P_i unilateral deviates from γ_i to $\bar{\gamma}_i$, then the *sign* of the change in their outcome is equal to the *sign* of the change in the potential, which is common to all players.

Notation. The rational for the name "potential" will become clear shortly.

and ϕ is called an *(exact) potential* for the game. A game is said to be an *ordinal potential game* if there exists a function $\phi(\gamma_1, \gamma_2, \ldots, \gamma_N)$ such that

$$J_i(\gamma_i, \gamma_{-i}) - J_i(\bar{\gamma}_i, \gamma_{-i}) > 0 \quad \Leftrightarrow \quad \phi(\gamma_i, \gamma_{-i}) - \phi(\bar{\gamma}_i, \gamma_{-i}) > 0,$$
$$\forall \gamma_i, \bar{\gamma}_i \in \Gamma_i, \quad \gamma_{-i} \in \Gamma_{-i}, \quad i \in \{1, 2, \ldots, N\} \tag{12.4}$$

and ϕ is called an *ordinal potential* for the game.

The exact or ordinal potentials of a game are not uniquely defined. For example, if $\phi(\cdot)$ is an exact/ordinal potential then, for every constant c, $\phi(\cdot) + c$ is also an exact/ordinal potential for the same game. It turns out that, for *exact potential games*, while the potential is not unique, all potentials differ only by an additive constant, i.e., if ϕ and $\bar{\phi}$ are both potentials for the same exact potential game, then there must exist a constant c such that

$$\phi(\gamma) = \bar{\phi}(\gamma) + c, \quad \forall \gamma \in \Gamma.$$

We refer the reader to [9, lemma 2.7] for a proof of this result.

12.2.1 Minima vs. Nash Equilibria in Potential Games

It turns out that for potential games, directionally-local minima of the potential ϕ are Nash equilibria:

Note. Recalling the Definition 9.5 of best-response equivalence, Proposition 12.2 could also be stated as "an exact (or ordinal) potential game with exact (or ordinal) potential ϕ is best-response equivalent to an identical interests game with outcomes given by ϕ." ▷ p. 111

Proposition 12.2. Consider an exact or ordinal potential game with (exact or ordinal) potential ϕ. An N-tuple of policies $\gamma^* := (\gamma_1^*, \gamma_2^*, \ldots, \gamma_N^*) \in \Gamma$ is a Nash equilibrium if and only if γ^* is a directionally-local minimum of ϕ. □

Attention! When the action spaces are finite, a global minimum always exists and therefore a directionally-local minimum also exists. In this case, potential games always have at least one Nash equilibrium.

Note. The implication that directionally-local minima of ϕ result in Nash equilibria only requires that $\phi(\gamma_i^*, \gamma_{-i}^*) - \phi(\gamma_i, \gamma_{-i}^*) \leq 0 \Rightarrow J_i(\gamma_i^*, \gamma_{-i}^*) - J_i(\gamma_i, \gamma_{-i}^*) \leq 0$. Games for which only this implication holds [instead of the equivalence in (12.4)] are called *generalized ordinal potential games.*

Proof of Proposition 12.2. Assuming that $\gamma^* := (\gamma_1^*, \gamma_2^*, \dots, \gamma_N^*)$ is a directionally-local minimum of ϕ, we have that

$$\phi(\gamma_i^*, \gamma_{-i}^*) - \phi(\gamma_i, \gamma_{-i}^*) \leq 0, \quad \forall \gamma_i \in \Gamma_i, \ i \in \{1, 2, \dots, N\}. \tag{12.5}$$

But then, both for exact and ordinal games, we also have that

$$J_i(\gamma_i^*, \gamma_{-i}^*) - J_i(\gamma_i, \gamma_{-i}^*) \leq 0, \quad \forall \gamma_i \in \Gamma_i, \ i \in \{1, 2, \dots, N\}, \tag{12.6}$$

which shows that γ^* is indeed a Nash equilibrium.

Conversely, if $\gamma^* := (\gamma_1^*, \gamma_2^*, \dots, \gamma_N^*)$ is a Nash equilibrium, then (12.6) holds, which for both exact and ordinal games, implies that (12.5) also holds, from which we conclude that $\gamma^* := (\gamma_1^*, \gamma_2^*, \dots, \gamma_N^*)$ is a directionally-local minimum of ϕ. ∎

Attention! For potential games we have an equivalence between directionally-local minima and Nash equilibria, but there is no match between global minima and admissible Nash equilibria. To verify this, consider a pure bimatrix game defined by the matrices

$$A = [a_{ij}]_{2\times 2} = \begin{bmatrix} \alpha_1 & \alpha_2 + 1 \\ \alpha_1 + 1 & \alpha_2 \end{bmatrix}, \qquad B = [b_{ij}]_{2\times 2} = \begin{bmatrix} \beta_1 & \beta_1 + 1 \\ \beta_2 + 1 & \beta_2 \end{bmatrix}.$$

One can verify that this is an exact potential game with potential given by

$$\Phi = [\phi_{ij}]_{2\times 2} = \begin{bmatrix} 0 & 1 \\ 1 & 0 \end{bmatrix}.$$

Since the potential Φ has two global minima, we necessarily have two Nash equilibria:

$$(1, 1) \text{ with outcomes } (\alpha_1, \beta_1)$$
$$(2, 2) \text{ with outcomes } (\alpha_2, \beta_2).$$

By appropriate choices of the constants $\alpha_1, \alpha_2, \beta_1, \beta_2$ we can make only one or both of these Nash equilibria admissible. This shows that global minima of the potential may not generate admissible Nash equilibria.

12.2.2 Bimatrix Potential Games

According to the previous definition of potential games, a pure bimatrix game defined by the two $m \times n$ matrices

$$A = [\, a_{ij} \,]_{m\times n}, \qquad B = [\, b_{ij} \,]_{m\times n},$$

is an (exact) potential game, if there exists a potential

$$\Phi = [\, \phi_{ij} \,]_{m\times n}$$

such that

$$a_{ij} - a_{\bar{i}j} = \phi_{ij} - \phi_{\bar{i}j}, \quad \forall i, \bar{i} \in \{1, 2, \dots, m\}, \ j \in \{1, 2, \dots, n\} \tag{12.7a}$$

$$b_{ij} - b_{i\bar{j}} = \phi_{ij} - \phi_{i\bar{j}}, \quad \forall i \in \{1, 2, \dots, m\}, \ j, \bar{j} \in \{1, 2, \dots, n\}. \tag{12.7b}$$

To verify whether or not such a potential exists, we can regard (12.7) as a linear system of equations with

$$\frac{m(m-1)}{2}n + m\frac{n(n-1)}{2}$$

equations and mn unknowns (the entries of Φ). When these equations have a solution, we can conclude that we have an exact potential game in pure policies.

Note. See Exercises 12.1 and 12.2. ▷ p. 142

It turns out that the equalities (12.7) also guarantee that the bimatrix game is an (exact) potential game in mixed policies:

Note. The generalization of this result for N-player games in normal form is tedious, but straightforward.

Proposition 12.3 (Potential bimatrix games). A bimatrix game defined by the two $m \times n$ matrices $A = [a_{ij}]$ and $B = [b_{ij}]$ is an exact potential game in pure or mixed policies if and only if there exists an $m \times n$ matrix $\Phi = [\phi_{ij}]$ for which (12.7) holds. □

Proof of Proposition 12.3. The proof of this result for the case of pure bimatrix games is simply a consequence of the fact that the conditions in (12.7) correspond precisely to the definition of an (exact) potential game.

For mixed policies, the conditions in (12.7) must be necessary since pure policies are special cases of mixed policies. Therefore, even when we are interested in mixed policies, the equalities in (12.3) must hold for all pure policies.

To prove that the conditions in (12.7) are also sufficient, we show next that $y'\Phi z$ is a potential for the mixed bimatrix game. Specifically, that given arbitrary mixed policies $y, \bar{y}$ for P_1 and $z, \bar{z}$ for P_2, we have that

$$y'Az - \bar{y}Az = y'\Phi z - \bar{y}\Phi z$$
$$y'Bz - yB\bar{z} = y'\Phi z - y\Phi\bar{z}.$$

To this effect, we start by expanding

$$y'Az - \bar{y}Az = \sum_{i=1}^{m}\sum_{j=1}^{n}(y_i - \bar{y}_i)a_{ij}z_j, \qquad y'Bz - yB\bar{z} = \sum_{i=1}^{m}\sum_{j=1}^{n} y_i b_{ij}(z_j - \bar{z}_j).$$

Because of (12.7), we conclude that for every $\bar{i}$ and $\bar{j}$, these differences must equal

$$\begin{aligned} y'Az - \bar{y}Az &= \sum_{i=1}^{m}\sum_{j=1}^{n}(y_i - \bar{y}_i)(a_{\bar{i}j} + \phi_{ij} - \phi_{\bar{i}j})z_j \\ &= \sum_{i=1}^{m}\sum_{j=1}^{n}(y_i - \bar{y}_i)\phi_{ij}z_j + \Big(\sum_{i=1}^{m}(y_i - \bar{y}_i)\Big)\Big(\sum_{j=1}^{n}(a_{\bar{i}j} - \phi_{\bar{i}j})z_j\Big) \\ y'Bz - yB\bar{z} &= \sum_{i=1}^{m}\sum_{j=1}^{n} y_i(b_{i\bar{j}} + \phi_{ij} - \phi_{i\bar{j}})(z_j - \bar{z}_j) \\ &= \sum_{i=1}^{m}\sum_{j=1}^{n} y_i\phi_{ij}(z_j - \bar{z}_j) + \Big(\sum_{i=1}^{m} y_i(b_{i\bar{j}} - \phi_{i\bar{j}})\Big)\Big(\sum_{j=1}^{n}(z_j - \bar{z}_j)\Big). \end{aligned}$$

But since $\sum_{j=1}^{n} y_i = \sum_{j=1}^{n} \bar{y}_i = \sum_{j=1}^{n} z_j = \sum_{j=1}^{n} \bar{z}_j = 1$, we conclude that $\sum_{i=1}^{m}(y_i - \bar{y}_i) = 0$ and $\sum_{j=1}^{n}(z_j - \bar{z}_j) = 0$ and therefore

$$y'Az - \bar{y}Az = \sum_{i=1}^{m}\sum_{j=1}^{n}(y_i - \bar{y}_i)\phi_{ij}z_j = y'\Phi z - \bar{y}\Phi z$$

$$y'Bz - yB\bar{z} = \sum_{i=1}^{m}\sum_{j=1}^{n} y_i\phi_{ij}(z_j - \bar{z}_j) = y'\Phi z - y\Phi\bar{z},$$

which concludes the sufficiency proof. ■

12.3 Characterization of Potential Games

The class of potential games is larger than one might think at first. *Identical interests games*, i.e., games for which there exists a function $\phi(\gamma)$ such that

$$J_i(\gamma) = \phi(\gamma), \quad \forall \gamma \in \Gamma, \quad i \in \{1, 2, \ldots, N\},$$

trivially satisfy (12.3) and therefore are potential games with potential ϕ.

Dummy games are games for which the outcome J_i of each player P_i does not depend on the player's own policy γ_i (but may depend on the policies of the other players), i.e.,

$$J_i(\gamma_i, \gamma_{-i}) = J_i(\bar{\gamma}_i, \gamma_{-i}) = J_i(\gamma_{-i}), \quad \forall \gamma_i, \bar{\gamma}_i \in \Gamma_i, \quad \gamma_{-i} \in \Gamma_{-i}, \quad i \in \{1, 2, \ldots, N\}.$$

Notation. For dummy games, we typically write the outcome of player P_i simply as $J_i(\gamma_{-i})$ to emphasize that it may depend on the policies γ_{-i} of the other players, but not on the policy γ_i of player P_i.

Notation. The terminology "dummy games" is motivated by the fact that for these games every N-tuple of policies is a Nash equilibrium.

Such games are also potential games with the constant potential

$$\phi(\gamma) = 0, \quad \forall \gamma \in \Gamma.$$

It turns out that the set of potential games is closed under "summation." To understand what is meant by this, suppose that one has two games G and H with the same *action spaces* $\Gamma_1, \Gamma_2, \ldots, \Gamma_N$ but different outcomes for the same set of policies $\gamma \in \Gamma$:

- G has an *outcome* given by $G_i(\gamma)$ for each player P_i, $\forall i \in \{1, 2, \ldots, N\}$.
- H has an *outcome* given by $H_i(\gamma)$ for each player P_i, $\forall i \in \{1, 2, \ldots, N\}$.

The *sum game* $G + H$ is a game with the same action spaces as G and H, but with the outcome

$$G_i(\gamma) + H_i(\gamma), \quad \forall \gamma \in \Gamma,$$

for each player P_i, $\forall i \in \{1, 2, \ldots, N\}$. The following result is straightforward to verify.

Proposition 12.4 (Sum of potential games). If G and H are two potential games with the same action spaces and potentials ϕ_G and ϕ_H, respectively, then the sum game $G + H$ is also a potential game with a potential $\phi_G + \phi_H$. □

An immediate corollary of this proposition is that the sum of an identical interests game and a dummy game is always a potential game. It turns out that any potential game can be expressed as the sum of two games of these types:

Proposition 12.5. A game G is an exact potential game if and only if there exists a dummy game D and an identical interests game H such that $G = D + H$. □

Proof of Proposition 12.5. Because of Proposition 12.4, we already knew that if G can be decomposed as the sum of a dummy game D and an identical interests game H then G must be a potential game.

To prove the converse, we need to show that if G is an exact potential game with potential ϕ_G, then we can find a dummy game D and an identical interests game H such that $G = D + H$. It turns out that this is simple. For the identical interests game H, we chose outcomes for all players equal to the potential ϕ_G:

$$H_i(\gamma) = \phi_G(\gamma), \quad \forall \gamma \in \Gamma.$$

This then uniquely defines what the outcomes of the dummy game D must be so that $G = D + H$:

$$D_i(\gamma) = G_i(\gamma) - \phi_G(\gamma), \quad \forall \gamma \in \Gamma.$$

To check that this indeed defines a dummy game, we compute

$$D_i(\gamma_i, \gamma_{-i}) - D_i(\bar{\gamma}_i, \gamma_{-i}) = G_i(\gamma) - G_i(\bar{\gamma}) - \phi_G(\gamma) + \phi_G(\bar{\gamma}),$$
$$\forall \gamma_i, \bar{\gamma}_i \in \Gamma_i, \ \gamma_{-i} \in \Gamma_{-i}, \ i \in \{1, 2, \dots, N\},$$

which is equal to zero because ϕ_G is an exact potential for G. This confirms that $D_i(\gamma_i, \gamma_{-i})$ indeed does not depend on γ_i. ■

12.4 Potential Games with Interval Action Spaces

We have just seen that constructing a potential game is very simple, because one only needs to add an identical interests game to a dummy game. It turns out that for games whose action spaces are intervals in the real line, determining if a given game is an exact potential game is also straightforward:

Note. A matrix game with mixed policies and only two actions is an example of a game whose action spaces are intervals.

Lemma 12.1 (Potential games with interval action spaces). For a given game G, suppose that every action space Γ_i is a close interval in $\mathbb{R}$ and every outcome J_i is twice continuously differentiable. In this case, the following three statements are equivalent:

1. G is an exact potential game.
2. There exists a twice differentiable function $\phi(\gamma)$ such that

$$\frac{\partial J_i(\gamma)}{\partial \gamma_i} = \frac{\partial \phi(\gamma)}{\partial \gamma_i}, \quad \forall \gamma \in \Gamma, \ i \in \{1, 2, \dots, N\} \tag{12.8}$$

and when these equalities hold ϕ is an exact potential for the game.

Notation 4. This property justifies the use of the terminology *potential* game. ▷ p. 141

3. The outcomes satisfy

$$\frac{\partial^2 J_i(\gamma)}{\partial \gamma_i \gamma_j} = \frac{\partial^2 J_j(\gamma)}{\partial \gamma_i \gamma_j}, \quad \forall \gamma \in \Gamma, \; i, j \in \{1, 2, \ldots, N\} \tag{12.9}$$

and when these equalities hold, we can construct an exact potential using

$$\phi(\gamma) = \sum_{k=1}^{N} \int_0^1 \frac{\partial J_k(\zeta + \tau(\gamma - \zeta))}{\partial \gamma_k} (\gamma_k - \zeta_k) d\tau, \quad \forall \gamma \in \Gamma, \tag{12.10}$$

where ζ can be any element of Γ. □

Note. To prove that multiple statements $P_1, P_2, \ldots, P_\ell$ are equivalent, one simply needs to prove a cycle of implications: $P_1 \Rightarrow P_2$, $P_2 \Rightarrow P_3, \ldots, P_{\ell-1} \Rightarrow P_\ell$, and $P_\ell \Rightarrow P_1$. Alternatively, we can also show pairwise equivalences: $P_1 \Rightarrow P_2$, $P_2 \Rightarrow P_1$, and $P_2 \Rightarrow P_3$, $P_3 \Rightarrow P_2$, etc. The latter approach is followed in the proof of Lemma 12.1.

Proof of Lemma 12.1. To prove that 1 implies 2 we assume that G is an exact potential game with potential ϕ. In this case, for every $\gamma_i, \bar{\gamma}_i, \gamma_{-i}$,

$$J_i(\gamma_i, \gamma_{-i}) - J_i(\bar{\gamma}_i, \gamma_{-i}) = \phi(\gamma_i, \gamma_{-i}) - \phi(\bar{\gamma}_i, \gamma_{-i}).$$

Dividing both sides by $\gamma_i - \bar{\gamma}_i$ and making $\bar{\gamma}_i \to \gamma_i$ we obtain precisely (12.8).

To show that 2 implies 1, we integrate both sides of (12.8) between $\bar{\gamma}_i$ and $\hat{\gamma}_i$ and obtain

$$\int_{\bar{\gamma}_i}^{\hat{\gamma}_i} \frac{\partial J_i(\gamma)}{\partial \gamma_i} d\gamma_i = \int_{\bar{\gamma}_i}^{\hat{\gamma}_i} \frac{\partial \phi(\gamma)}{\partial \gamma_i} d\gamma_i, \quad \forall \bar{\gamma}_i, \hat{\gamma}_i, \gamma_{-i}, \; i \in \{1, 2, \ldots, N\},$$

which is equivalent to

$$J_i(\hat{\gamma}_i, \gamma_{-i}) - J_i(\bar{\gamma}_i, \gamma_{-i}) = \phi(\hat{\gamma}_i, \gamma_{-i}) - \phi(\bar{\gamma}_i, \gamma_{-i}), \quad \forall \bar{\gamma}_i, \hat{\gamma}_i, \gamma_{-i}, \; i \in \{1, 2, \ldots, N\},$$

proving that we have a potential game with potential ϕ. At this point we have shown that 1 and 2 are equivalent.

To prove that 2 implies 3, we take partial derivatives of both sides of (12.8) with respect to γ_j and conclude that

$$\frac{\partial^2 J_i(\gamma)}{\partial \gamma_i \gamma_j} = \frac{\partial^2 \phi(\gamma)}{\partial \gamma_i \gamma_j}, \quad \forall \gamma, \; i \in \{1, 2, \ldots, N\}.$$

Since the right-hand side of the above equality does not change if we exchange i by j, we immediately conclude that the left-hand side cannot change if we exchange i by j. This is precisely, what is stated in 3.

To show that 3 implies 2, we will show that the function defined in (12.10) satisfies (12.8). To this effect, we need to compute

$$\frac{\partial \phi(\gamma)}{\partial \gamma_i} = \frac{\partial}{\partial \gamma_i}\left(\int_0^1 \frac{\partial J_i(\zeta + \tau(\gamma - \zeta))}{\partial \gamma_i}(\gamma_i - \zeta_i)d\tau \right.$$

$$\left. + \sum_{k \neq i} \int_0^1 \frac{\partial J_k(\zeta + \tau(\gamma - \zeta))}{\partial \gamma_k}(\gamma_k - \zeta_k)d\tau \right)$$

$$= \int_0^1 \tau \frac{\partial^2 J_i(\zeta + \tau(\gamma - \zeta))}{\partial \gamma_i^2}(\gamma_i - \zeta_i)d\tau + \int_0^1 \frac{\partial J_i(\zeta + \tau(\gamma - \zeta))}{\partial \gamma_i}d\tau$$

$$+ \sum_{k \neq i} \int_0^1 \tau \frac{\partial^2 J_k(\zeta + \tau(\gamma - \zeta))}{\partial \gamma_i \gamma_k}(\gamma_k - \zeta_k)d\tau$$

$$= \int_0^1 \frac{\partial J_i(\zeta + \tau(\gamma - \zeta))}{\partial \gamma_i}d\tau + \sum_{k=1}^{N} \int_0^1 \tau \frac{\partial^2 J_k(\zeta + \tau(\gamma - \zeta))}{\partial \gamma_k \gamma_i}(\gamma_k - \zeta_k)d\tau.$$

Using (12.9), we conclude that

$$\frac{\partial \phi(\gamma)}{\partial \gamma_i} = \int_0^1 \frac{\partial J_i(\zeta + \tau(\gamma - \zeta))}{\partial \gamma_i}d\tau + \int_0^1 \tau \sum_{k=1}^{N} \frac{\partial^2 J_i(\zeta + \tau(\gamma - \zeta))}{\partial \gamma_k \gamma_i}(\gamma_k - \zeta_k)d\tau$$

$$= \int_0^1 \frac{\partial J_i(\zeta + \tau(\gamma - \zeta))}{\partial \gamma_i}d\tau$$

$$+ \int_0^1 \tau \left(\sum_{k=1}^{N} \frac{\partial}{\partial \gamma_k} \frac{\partial J_i(\zeta + \tau(\gamma - \zeta))}{\partial \gamma_i} \frac{d(\zeta_k + \tau(\gamma_k - \zeta_k))}{d\tau} \right) d\tau$$

$$= \int_0^1 \frac{\partial J_i(\zeta + \tau(\gamma - \zeta))}{\partial \gamma_i}d\tau + \int_0^1 \tau \frac{d}{d\tau} \frac{\partial J_i(\zeta + \tau(\gamma - \zeta))}{\partial \gamma_i}d\tau.$$

Integrating by parts, we finally obtain

$$\frac{\partial \phi(\gamma)}{\partial \gamma_i} = \int_0^1 \frac{\partial J_i(\zeta + \tau(\gamma - \zeta))}{\partial \gamma_i}d\tau + [\tau \frac{\partial J_i(\zeta + \tau(\gamma - \zeta))}{\partial \gamma_i}]_0^1$$

$$- \int_0^1 \frac{\partial J_i(\zeta + \tau(\gamma - \zeta))}{\partial \gamma_i}d\tau = \frac{\partial J_i(\gamma))}{\partial \gamma_i},$$

which proves 2. At this point we have also shown that 2 and 3 are equivalent, which completes the proof. ■

Notation 4 (Potential games). This property justifies the use of the terminology *potential* game: If we view the (symmetric of the) vector of the derivatives of the outcomes J_i with respect to the corresponding γ_i as a "force"

$$F := -\left(\frac{\partial J_1}{\partial \gamma_1}, \frac{\partial J_2}{\partial \gamma_2}, \ldots, \frac{\partial J_N}{\partial \gamma_N}\right)$$

that drives the players towards a (selfish) minimization of their outcomes, then condition 2 corresponds to requirement that this force be conservative with a potential ϕ. Recall that a mechanical force F is *conservative* if it can be written as $F = -\nabla\phi$ for some potential ϕ. A mechanically inclined reader may construct potential games with outcomes inspired by conservative forces. □

12.5 Practice Exercises

12.1. Consider a bimatrix game with two actions for each player defined by the following matrices

$$A = \underbrace{\begin{bmatrix} a_{11} & a_{12} \\ a_{21} & a_{22} \end{bmatrix}}_{\text{P}_2 \text{ choices}} \Big\} \text{P}_1 \text{ choices} \qquad B = \underbrace{\begin{bmatrix} b_{11} & b_{12} \\ b_{21} & b_{22} \end{bmatrix}}_{\text{P}_2 \text{ choices}} \Big\} \text{P}_1 \text{ choices}.$$

1. Under what conditions is this an exact potential game in *pure* policies?

   Your answer should be a set of equalities/inequalities that the a_{ij} and b_{ij} need to satisfy.

2. Show that the prisoners' dilemma bimatrix game defined by the following matrices

$$A = \underbrace{\begin{bmatrix} 2 & 30 \\ 0 & 8 \end{bmatrix}}_{\text{P}_2 \text{ choices}} \Big\} \text{P}_1 \text{ choices} \qquad B = \underbrace{\begin{bmatrix} 2 & 0 \\ 30 & 8 \end{bmatrix}}_{\text{P}_2 \text{ choices}} \Big\} \text{P}_1 \text{ choices}$$

   is an exact potential game in pure policies. □

Solution to 12.1.

1. For this game to be a potential game, with potential

$$\begin{bmatrix} \phi_{11} & \phi_{12} \\ \phi_{21} & \phi_{22} \end{bmatrix}$$

   we need to have

$$a_{11} - a_{21} = \phi_{11} - \phi_{21} \qquad a_{12} - a_{22} = \phi_{12} - \phi_{22},$$
$$b_{11} - b_{12} = \phi_{11} - \phi_{12} \qquad b_{21} - b_{22} = \phi_{21} - \phi_{22},$$

   which is equivalent to

$$a_{11} - a_{21} = \phi_{11} - \phi_{21} \qquad a_{12} - a_{22} = \phi_{12} - \phi_{22},$$
$$a_{11} - a_{21} - (b_{11} - b_{12}) = \phi_{12} - \phi_{21}, \quad a_{12} - a_{22} - (b_{21} - b_{22}) = \phi_{12} - \phi_{21}.$$

   This system of equations has a solution if and only if

$$a_{11} - a_{21} - (b_{11} - b_{12}) = a_{12} - a_{22} - (b_{21} - b_{22}),$$

   in which case we can make, e.g.,

$$\phi_{22} = 0, \qquad \phi_{12} = a_{12} - a_{22},$$
$$\phi_{21} = \phi_{12} - a_{11} + a_{21} + b_{11} - b_{12}, \qquad \phi_{11} = \phi_{21} + a_{11} - a_{21}.$$

2. The potential for this game can be defined by the following matrix

$$\Phi = \underbrace{\begin{bmatrix} 24 & 22 \\ 22 & 0 \end{bmatrix}}_{\text{P}_2 \text{ choices}} \Big\} \text{P}_1 \text{ choices}.$$

Indeed,

$$a_{11} - a_{21} = \phi_{11} - \phi_{21} = 2 \qquad a_{12} - a_{22} = \phi_{12} - \phi_{22} = 22,$$
$$b_{11} - b_{12} = \phi_{11} - \phi_{12} = 2 \qquad b_{21} - b_{22} = \phi_{21} - \phi_{22} = 22. \qquad \square$$

12.2. Consider a bimatrix game with two actions for both players defined by the following matrices

$$A = \underbrace{\begin{bmatrix} a_{11} & a_{12} \\ a_{21} & a_{22} \end{bmatrix}}_{\mathsf{P}_2 \text{ choices}} \Big\} \mathsf{P}_1 \text{ choices} \qquad B = \underbrace{\begin{bmatrix} b_{11} & b_{12} \\ b_{21} & b_{22} \end{bmatrix}}_{\mathsf{P}_2 \text{ choices}} \Big\} \mathsf{P}_1 \text{ choices}.$$

1. Under what conditions is this an exact potential game in *mixed* policies?

   Your answer should be a set of equalities/inequalities that the a_{ij} and b_{ij} need to satisfy.

2. Find an exact potential function when the game is a potential game.

   Your answer should be a function that depends on the a_{ij} and b_{ij}.

3. When is a zero-sum game an exact potential game in mixed policies? Find its potential function.

   Your answer should be a set of equalities/inequalities that the a_{ij} and b_{ij} need to satisfy and the potential function should be a function that depends on the a_{ij} and b_{ij}. □

Solution to 12.2.

1. Under the mixed policies

$$y := \begin{bmatrix} y_1 \\ 1 - y_1 \end{bmatrix}, \qquad y_1 \in [0, 1] \qquad z := \begin{bmatrix} z_1 \\ 1 - z_1 \end{bmatrix}, \qquad z_1 \in [0, 1],$$

for P_1 and P_2, respectively, the outcomes for this games are given by

$$J_1(y_1, z_1) = a_{11} y_1 z_1 + a_{12} y_1 (1 - z_1) + a_{21}(1 - y_1) z_1 + a_{22}(1 - y_1)(1 - z_1)$$
$$J_2(y_1, z_1) = b_{11} y_1 z_1 + b_{12} y_1 (1 - z_1) + b_{21}(1 - y_1) z_1 + b_{22}(1 - y_1)(1 - z_1)$$

and therefore

$$\frac{\partial^2 J_1}{\partial y_1 z_1} = a_{11} - a_{12} - a_{21} + a_{22} \qquad \frac{\partial^2 J_2}{\partial y_1 z_1} = b_{11} - b_{12} - b_{21} + b_{22}$$
$$\frac{\partial^2 J_1}{\partial y_1^2} = 0 \qquad \frac{\partial^2 J_2}{\partial y_1^2} = 0$$
$$\frac{\partial^2 J_1}{\partial z_1^2} = 0 \qquad \frac{\partial^2 J_2}{\partial z_1^2} = 0.$$

In view of Proposition 12.1, we conclude that this is a potential game if and only if

$$a_{11} - a_{12} - a_{21} + a_{22} = b_{11} - b_{12} - b_{21} + b_{22}. \tag{12.11}$$

Note. Note that the prisoners' dilemma bimatrix game in Example 12.1 satisfies this condition.

2. Still according to Proposition 12.1, the potential function can be obtained by

$$\begin{aligned}
\phi(y_1, z_1) &= \int_0^1 \frac{\partial J_1(\tau y_1, \tau z_1)}{\partial y_1} y_1 + \frac{\partial J_2(\tau y_1, \tau z_1)}{\partial z_1} z_1 d\tau \\
&= \int_0^1 \big(a_{11}\tau z_1 + a_{12}(1-\tau z_1) - a_{21}\tau z_1 - a_{22}(1-\tau z_1)\big) y_1 \\
&\quad + \big(b_{11}\tau y_1 - b_{12}\tau y_1 + b_{21}(1-\tau y_1) - b_{22}(1-\tau y_1)\big) z_1 d\tau \\
&= \int_0^1 \big((a_{11} - a_{12} - a_{21} + a_{22})\tau z_1 + a_{12} - a_{22}\big) y_1 \\
&\quad + \big((b_{11} - b_{12} - b_{21} + b_{22})\tau y_1 + b_{21} - b_{22}\big) z_1 d\tau \\
&= \frac{1}{2}(a_{11} - a_{12} - a_{21} + a_{22}) y_1 z_1 + (a_{12} - a_{22}) y_1 \\
&\quad + \frac{1}{2}(b_{11} - b_{12} - b_{21} + b_{22}) y_1 z_1 + (b_{21} - b_{22}) z_1.
\end{aligned}$$

Note. Note that this is consistent with what we saw in Exercise 12.1.

Using (12.11), this simplifies to

$$\phi(y_1, z_1) = (a_{11} - a_{12} - a_{21} + a_{22}) y_1 z_1 + (a_{12} - a_{22}) y_1 + (b_{21} - b_{22}) z_1.$$

3. For a zero-sum game the left and the right hand sides of (12.11) are symmetric, which is only possible if

$$a_{11} - a_{12} - a_{21} + a_{22} = 0 \Leftrightarrow a_{11} + a_{22} = a_{12} + a_{21}$$

which corresponds to the outcomes

$$J_1(y_1, z_1) = a_{12} y_1 + a_{21} z_1 + a_{22} - a_{22} y_1 - a_{22} z_1 = -J_2(y_1, z_1).$$

In this case, a potential function is given by

$$\phi(y_1, z_1) = (a_{12} - a_{22}) y_1 - (a_{21} - a_{22}) z_1 = (a_{12} - a_{22})(y_1 - z_1). \quad \square$$

12.6 Additional Exercise

12.3. Prove Proposition 12.4.

LECTURE 13

Classes of Potential Games

In spite of the fact that every potential game must be the sum of an identical interests game and a dummy game, potential games can be used to capture a wide range of problems. In this lecture, we consider several classes of potential games that are common in the literature and discuss how one can obtain Nash equilibria for such games.

13.1 Identical Interests Plus Dummy Games

Suppose that we can write the outcomes J_i of a game G as

$$J_i(\gamma_i, \gamma_{-i}) = \phi(\gamma_i, \gamma_{-i}) + Q_i(\gamma_{-i}), \quad \forall \gamma_i \in \Gamma_i, \ \ \gamma_{-i} \in \Gamma_{-i}, \ \ i \in \{1, 2, \ldots, N\} \tag{13.1}$$

for appropriate functions $\phi(\gamma)$, $Q_i(\gamma_{-i})$, $i \in \{1, 2, \ldots, N\}$. In this case, G is the sum of an identical interests game with outcomes $\phi(\gamma)$ and a dummy game with outcomes $Q_i(\gamma_{-i})$. In view of Proposition 12.5, we conclude that G is an exact potential game and ϕ is an exact potential for G.

While we have seen that every potential game can be decomposed as in (13.1), we shall see next several other constructions that also lead to potential games, but arise more naturally in the context of specific applications.

13.2 Decoupled Plus Dummy Games

Notation. For decoupled games, we typically write the outcome of player P_i simply as $H_i(\gamma_i)$ to emphasize that it depends on the policy γ_i of player P_i, but not on the policies γ_{-i} of the remaining players.

A game H with outcomes H_i is said to be a *decoupled game* if the outcome for player P_i only depends on P_1's own policy γ_i, i.e., if

$$H_i(\gamma_i, \gamma_{-i}) = H_i(\gamma_i, \bar{\gamma}_{-i}) = H_i(\gamma_i), \quad \forall \gamma_i \in \Gamma_i, \ \gamma_{-i}, \bar{\gamma}_{-i} \in \Gamma_{-i}, \ i \in \{1, 2, \dots, N\}. \tag{13.2}$$

Decoupled games are potential games with potential

Note. See Exercise 13.1.
▷ p. 159

$$\phi(\gamma) = \sum_{i=1}^{N} H_i(\gamma_i), \quad \forall \gamma_i \in \Gamma_i. \tag{13.3}$$

Suppose that we can write the outcomes J_i of a game G as

$$J_i(\gamma_i, \gamma_{-i}) = H_i(\gamma_i) + Q_i(\gamma_{-i}), \quad \forall \gamma_i \in \Gamma_i, \ \gamma_{-i} \in \Gamma_{-i}, \ i \in \{1, 2, \dots, N\} \tag{13.4}$$

for appropriate functions $H_i(\gamma_i)$, $Q_i(\gamma_{-i})$, $i \in \{1, 2, \dots, N\}$. In this case, G is the sum of a decoupled game with outcomes $H_i(\gamma_i)$ and a dummy game with outcomes $Q_i(\gamma_{-i})$. In view of Proposition 12.4, we conclude that G is an exact potential game and, since any dummy game admits a zero potential, we conclude that (13.3) is also a potential for G.

Note. See Exercise 13.7.
▷ p. 167

Example 13.1 (Wireless power control game). Suppose that N players want to transmit data through a wireless medium and have to decide how much power to use in their transmitters. Each player P_i must select a power level γ_i within a (perhaps finite) set of admissible power levels:

$$\gamma_i \in \Gamma_i := [p_{\min}, p_{\max}] \subset (0, \infty).$$

However, the transmission of each player appears as background noise for the remaining players. In particular, the player P_i observes a signal to interference plus noise ratio (SINR) given by

$$\frac{\gamma_i}{n + \alpha \sum_{j \neq i} \gamma_j},$$

where n denotes some constant background noise and $\alpha \in (0, 1)$ a gain that reflects how well the receiver can reject power from the signals that it is not trying to decode. The goal of each player P_i is to minimize a cost that is the sum of two terms. The first term is given by

Note. The motivation for the term in (13.5) stems from Shannon-Hartley's capacity Theorem, which states that the largest bit-rate that can be sent through an analog communication channel with bandwidth B, with a given average signal power S and subject to additive white Gaussian noise of power N, is given by $B \log_2(1 + \frac{S}{N}) \approx B \log_2(\frac{S}{N})$, where the approximate formula is valid for signal-to-noise ratios much larger than one [6].

$$-\beta \log\left(\frac{\gamma_i}{n + \alpha \sum_{j \neq i} \gamma_i}\right), \tag{13.5}$$

and expresses the fact that it is desirable to have a large SINR to increase the effective bit-rate; whereas the second term expresses the cost of utilizing a large amount of power and is given by

$$C(\gamma_i),$$

for some monotone non-decreasing *cost function*. The resulting outcome that P_i wants to minimize is therefore given by

$$J_i(\gamma) = C(\gamma_i) - \beta \log(\gamma_i) + \beta \log\left(n + \alpha \sum_{j \neq i} \gamma_j\right),$$

which is of the form (13.4) and therefore we have a potential game with potential

$$\phi(\gamma) := \sum_{i=1}^{N} \left(C(\gamma_i) - \beta \log(\gamma_i)\right). \tag{13.6}$$

This means that any global minimizer to (13.6) is a Nash equilibrium to this game.

Suppose now that the cost $C(\gamma_i)$ is controlled by a network administrator that charges the players by their use of power. For example, if this network administrator sets

$$C(s) := \beta \log(s) + s, \tag{13.7}$$

then the potential becomes

Note. For the cost C in (13.7), the global minimum of (13.8) corresponds to all players using the least amount of power. This particular potential does not have directionally-local minima other than the global minima so it has a unique Nash equilibrium (see Proposition 12.2). ▷ p. 135

$$\phi(\gamma) := \sum_{i=1}^{N} \gamma_i, \tag{13.8}$$

which is equal to the total power that will eventually interfere with other wireless users. In this case, by playing at Nash the players actually achieve the social optimum. Alternatively, the network administrator could set

$$C(s) := \beta^* \log(s),$$

where β^* can be viewed as the price for power (in dBs). This leads to the potential

Note. What would be achieved by selecting $C(s) := \beta \log(s) - s$ or by selecting $C(s) := \beta \log(s) - (s - s^*)^2$ for a given $s^* \in \mathbb{R}$?

$$\phi(\gamma) := (\beta^* - \beta) \sum_{i=1}^{N} \log(\gamma_i),$$

which is minimized for $\gamma_i = p_{\text{min}}$ or $\gamma_i = p_{\text{max}}$ depending on whether $\beta^* < \beta$ or $\beta^* > \beta$, respectively. By adjusting the price β^* and observing the players' reaction, the administrator could learn the value of the players' utility parameter β. By charging different costs, the network administrator can "force" the players to "selfishly" minimize a wide range of other costs. □

13.3 Bilateral Symmetric Games

A game G is called a *bilateral symmetric game* if its outcomes J_i are of the form

$$J_i(\gamma) = H_i(\gamma_i) + \sum_{j \neq i} W_{ij}(\gamma_i, \gamma_j), \quad \forall \gamma_i \in \Gamma_i, \quad i \in \{1, 2, \ldots, N\}, \tag{13.9}$$

for appropriate functions $H_i(\gamma_i)$, $W_{ij}(\gamma_i, \gamma_j)$, $i, j \in \{1, 2, \ldots, N\}$, with

$$W_{ij}(\gamma_i, \gamma_j) = W_{ji}(\gamma_j, \gamma_i), \quad \forall \gamma_i, \gamma_j \in \Gamma_i, \;\; i, j \in \{1, 2, \ldots, N\}. \tag{13.10}$$

We can view the outcome (13.9) as the sum of a "decoupled" term $H_i(\gamma_i)$ that only depends on the policy of player P_i and several terms $W_{ij}(\gamma_i, \gamma_j)$, $j \neq i$ that express *bilateral interactions* between player P_i and other players P_j, $j \neq i$. In bilateral symmetric games, these interactions must be "symmetric" in the sense of (13.10), i.e., the effect $W_{ij}(\gamma_i, \gamma_j)$ of player P_j on player P_i must be the same as the effect $W_{ji}(\gamma_j, \gamma_i)$ of player P_i on player P_j.

Note. See Exercise 13.3. ▷ p. 161

One can show that a bilateral symmetric game G is an exact potential game with the following potential:

$$\phi(\gamma) = \sum_{i=1}^{N} \left(H_i(\gamma_i) + \sum_{j=1}^{k-1} W_{ij}(\gamma_i, \gamma_j) \right), \quad \forall \gamma_i \in \Gamma_i. \tag{13.11}$$

13.4 Congestion Games

Suppose that N players must share R resources and that each player must decide which resources to use. A *policy* γ_i for player P_i is thus a subset of the R resources that the player wants to use, i.e.,

$$\gamma_i \subset \mathcal{R} := \{1, 2, \ldots, R\}.$$

The *action space* Γ_i for player P_i is the collection of all such sets that can be used to accomplish a particular goal for the player.

Notation. Given a set $\mathcal{A}$, we denote by $|\mathcal{A}|$ the cardinality (i.e., number of elements) of the set $\mathcal{A}$.

In a *congestion game*, the outcomes for the different players depend on the total number of players that use each resource $r \in \mathcal{R}$, which can be written as $|\mathcal{S}_r(\gamma_1, \gamma_2, \ldots, \gamma_N)|$, where

$$\mathcal{S}_r(\gamma_1, \gamma_2, \ldots, \gamma_N) := \left\{ i \in \{1, 2, \ldots, N\} : r \in \gamma_i \right\}$$

Note. In congestion games, all players are equal, in the sense that the cost associated with each resource only depends on the total number of players using that resource and not on which players use it.

denotes the set of players using the resource $r \in \mathcal{R}$. In particular, the *outcome* for player P_i is of the form

$$J_i(\gamma) := \sum_{r \in \gamma_i} C_r(|\mathcal{S}_r(\gamma)|), \gamma := (\gamma_1, \gamma_2, \ldots, \gamma_N), \tag{13.12}$$

where the summation is over every resource $r \in \mathcal{R}$ used by the policy γ_i of player P_i and the function $C_r(n)$ provides the cost to each player of using resource r when n players share this resource. Typically, $C_r(n)$ is a monotonically increasing function, expressing the fact that when many players use the same resources the resulting "congestion" increases the cost for everyone.

Note. See Exercise 13.10. ▷ p. 167

One can show that a congestion game is an exact potential game with the following potential:

$$\phi(\gamma) := \sum_{r \in \mathcal{R}} \sum_{k=1}^{|\mathcal{S}_r(\gamma)|} C_r(k), \quad \forall \gamma_i \in \Gamma_i \tag{13.13}$$

and therefore the Nash equilibria of this game will be the directionally-local minima of this function.

Example 13.2 (Homogeneous vehicle routing). Consider the following road network that connects the four cities A, B, C, D through five roads:

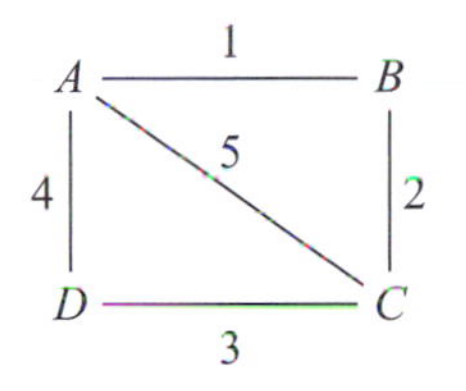

Assuming that player P_1 wants to route vehicles from city A to city D and player P_2 wants to route vehicles from city B to city D, their action spaces need to include all policies (i.e., sets of routes/resources) that allow them to route vehicles as desired. Specifically, the *action spaces* for players P_1 and P_2 are given, respectively, by

$$\Gamma_1 := \{\{4\}, \{5, 3\}, \{1, 2, 3\}\}, \qquad \Gamma_2 := \{\{1, 4\}, \{2, 3\}, \{1, 5, 3\}, \{2, 5, 4\}\},$$

Suppose that we take the cost $C_r(n)$ in (13.12) to be the time it takes to travel along the road $r \in \mathcal{R} := \{1, 2, 3, 4, 5\}$ when $n \in \{1, 2\}$ players use that road to route vehicles, with the understanding that if more vehicles use the same road they need to drive slower. In this case, the *outcome* $J_i(\gamma)$ for player P_i would be the time it takes for P_i's vehicles to go from their start to their end city. □

Note. For a congestion game, the time it takes to travel along a road must depend solely on how many players use that road to route vehicles and not on which specific players use it. This carries the implicit homogeneity assumption that all players are equal.

13.5 Other Potential Games

The class of potential games is very large and contains several unexpected examples. The *Sudoku* puzzle consists of filling a 9×9 grid with digits 1 through 9 so that each row, each column, and each of the nine 3×3 sub-grids (also called blocks) contain all nine digits without repetitions (see Figure 13.1).

5	3			7				
6			1	9	5			
	9	8					6	
8				6				3
4			8		3			1
7				2				6
	6					2	8	
			4	1	9			5
				8			7	9

5	3	4	6	7	8	9	1	2
6	7	2	1	9	5	3	4	8
1	9	8	3	4	2	5	6	7
8	5	9	7	6	1	4	2	3
4	2	6	8	5	3	7	9	1
7	1	3	9	2	4	8	5	6
9	6	1	5	3	7	2	8	4
2	8	7	4	1	9	6	3	5
3	4	5	2	8	6	1	7	9

Figure 13.1 Sudoku puzzle in its initial configuration (left) and solved (right).

Imagine that Sudoku is played by $N := 81$ players, each placed at one square of the 9×9 grid, with the understanding that the player at the ith square decides which digit to place in its own square. The players associated with squares that are originally filled have action spaces with a single element (the digit in their square), whereas the players associated with empty squares all have the same action space consisting of all 9 digits:

$$\Gamma_i := \{1, 2, 3, 4, 5, 6, 7, 8, 9\}.$$

Further suppose that the outcome of player P_i is given by

$$J_i(\gamma) := \sigma_i^{\text{row}}(\gamma) + \sigma_i^{\text{col}}(\gamma) + \sigma_i^{\text{block}}(\gamma),\ \gamma := \{\gamma_1, \gamma_2, \ldots, \gamma_N\},$$

where $\sigma_i^{\text{row}}(\gamma)$, $\sigma_i^{\text{col}}(\gamma)$, and $\sigma_i^{\text{block}}(\gamma)$ denote the number of times that the digit γ_i selected by P_i appears elsewhere in this player's row, column, block, respectively. For this selection of outcomes, the Sudoku puzzle is solved for every multiplayer policy γ that leads to a zero outcome for every player.

Note. See Exercise 13.2. ▷ p. 160

Note. It is possible to construct other multiplayer potential games that have Nash equilibria at the solution of the Sudoku puzzle. See Exercise 13.8. ▷ p. 167

Note. The results in the subsequent Section 13.7 will permit the construction of Sudoku solvers, based on the observation that solutions to Sudoku puzzles are Nash equilibria of potential games. See Exercises 13.5 and 13.9. ▷ p. 163

One can show that this multiplayer version of Sudoku is an exact potential game with potential

$$\phi(\gamma) := \frac{1}{2}\sum_{i=1}^{N} J_i(\gamma).$$

This potential function has a global minimum at 0 for every multiplayer policy γ that solves the Sudoku puzzle. However, one needs to be aware that there may be Nash equilibria that are directionally-local minima, but not global minima and thus do not solve the Sudoku puzzle.

This example illustrates how one can use noncooperative game theory to construct a simple distributed multi-agent algorithm to solve a problem that starts as a "single-agent" centralized optimization. For certain types of optimizations there are systematic procedures to construct multiplayer games whose Nash equilibria correspond to global minima; a topic that is explored in the next section.

13.6 Distributed Resource Allocation

Note. Distributed resource allocation problems do not start as noncooperative games, we emphasize this by calling the decision makers *agents* instead of *players*.

In a distributed resource allocation problem, N agents must share R resources and each agent must decide which resources to use. A *resource allocation policy* γ_i for agent A_i is thus a subset of the R resources that the agent wants to use, i.e.,

$$\gamma_i \subset \mathcal{R} := \{1, 2, \ldots, R\}.$$

The *action space* Γ_i for the agent A_i is the collection of all such sets that can be used to accomplish a particular goal for the agent.

Note. The criteria in resource allocation problems is more general than the one used in the (closely-related) congestion games, where the outcomes of the players were only allowed to depend on the number of players $\sigma_r(\gamma)$ using a specific resource r and not on which agents $\mathcal{S}_r(\gamma)$ use that specific resource.

In a *distributed resource allocation optimization*, the criteria depends on the sets of agents $\mathcal{S}_r(\gamma)$ that use each resource $r \in \mathcal{R}$, which are given by

$$\mathcal{S}_r(\gamma_1, \gamma_2, \ldots, \gamma_N) := \left\{i \in \{1, 2, \ldots, N\} : r \in \gamma_i\right\}.$$

The goal is to find a joint policy $\gamma = (\gamma_1, \gamma_2, \ldots, \gamma_N)$ for all the agents that minimizes a *global welfare cost* of the form

$$W(\gamma) := \sum_{r\in\mathcal{R}} W_r\left(\mathcal{S}_r(\gamma)\right), \tag{13.14}$$

where the function $W_r(\cdot)$ is called the *welfare cost for resource r* and maps each possible set of agents using resource r to a numerical value. Global welfare costs like (13.14), which are expressed by a sum of welfare costs for individual resources, are called *separable*.

Several examples of distributed allocation problems with separable global welfare costs can be found in [7, section 6], and include graph coloring problems and (sensor) coverage problems. In graph coloring problems one wants to assign colors to the edges of a graph, while avoiding neighbors to have the same color. We can view Sudoku as a very complex graph coloring problem.

Example 13.3 (Heterogeneous vehicle routing). Consider a variation of the vehicle routing problem considered in Example 13.2, with the same road network connecting the four cities A, B, C, D:

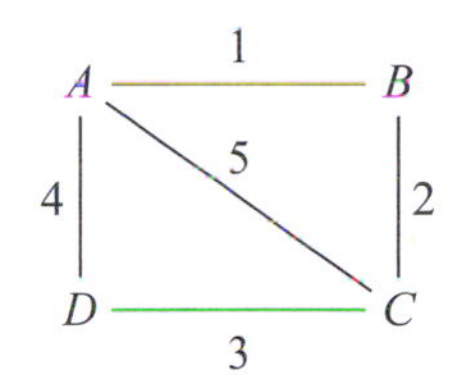

Assuming that agent A_1 wants to route vehicles from city A to city D and agent A_2 wants to route vehicles from city B to city D, their action spaces are, respectively, given by

$$\Gamma_1 := \{\{4\}, \{5, 3\}, \{1, 2, 3\}\}, \qquad \Gamma_2 := \{\{1, 4\}, \{2, 3\}, \{1, 5, 3\}, \{2, 5, 4\}\}.$$

Note. For distributed resource allocation, the time it takes to travel along a road may depend on which agents' vehicles travel that road, which means that one can consider agents that are heterogeneous in the way that they"congest" the road.

In a distributed resource allocation problem, one could seek for a joint policy $\gamma = (\gamma_1, \gamma_2)$ that minimizes the total time that the vehicles spend on their trips, which can be written as in (13.14), where $\mathcal{R} := \{1, 2, 3, 4, 5\}$; $\mathcal{S}_r(\gamma) \subset \{1, 2\}$ denotes the set of agents using road $r \in \mathcal{R}$ under the policy γ; and $W_r(\mathcal{S})$ the total time that the vehicles from the agents in the set $\mathcal{S} \subset \{1, 2\}$ spend on the road $r \in \mathcal{R}$, when these agents share that road. □

13.6.1 Distributed Welfare Games

While distributed resource allocation optimizations are typically not formulated as noncooperative games, it is possible to construct outcomes $J_i(\gamma)$ for the agents so that if they are encouraged to behave as players in the noncooperative game defined by these outcomes, the minimum of the global welfare cost (13.14) will correspond to a Nash equilibrium. Such games are called *distributed welfare games* and several options are possible for the construction of these games:

1. The *marginal contribution* or *wonderful life utility* corresponds to setting the outcome $J_i(\gamma)$ for each player P_i equal to

$$J_i(\gamma) := \sum_{r \in \gamma_i} \left(W_r(\mathcal{S}_r(\gamma)) - W_r(\mathcal{S}_r(\gamma) \setminus \{i\}) \right), \tag{13.15}$$

where the summation is taken over all resources r used by P_i's policy γ_i and each term accounts for the difference between the welfare cost associated with all players $\mathcal{S}_r(\gamma)$ using resource r and the cost associated with all but the ith player using the same resource.

The wonderful life utility leads to an exact potential game with potential equal to the global welfare cost

Note. See Exercise 13.4. ▷ p. 162

$$\phi(\gamma) := W(\gamma) = \sum_{r \in \mathcal{R}} W_r(\mathcal{S}_r(\gamma)), \tag{13.16}$$

whose minimum will be a Nash equilibrium.

2. The *Shapley value utility* corresponds to setting the outcome $J_i(\gamma)$ for each player P_i equal to

Notation. Given a set $\mathcal{A}$, we denote by $|\mathcal{A}|$ the cardinality (i.e., number of elements) of the set $\mathcal{A}$.

$$J_i(\gamma) := \sum_{r \in \gamma_i} \sum_{\mathcal{S} \subset \mathcal{S}_r(\gamma) \setminus \{i\}} \frac{|\mathcal{S}|!\,(|\mathcal{S}_r(\gamma)| - |\mathcal{S}| - 1)!}{|\mathcal{S}_r(\gamma)|!} \left(W_r(\mathcal{S} \cup \{i\}) - W_r(\mathcal{S}) \right), \tag{13.17}$$

where the summation is taken over all subsets $\mathcal{S}$ of $\mathcal{S}_r(\gamma) \setminus \{i\}$ and $W_r(\mathcal{S} \cup \{i\}) - W_r(\mathcal{S})$ compares the marginal cost of adding player P_i to the set of players $\mathcal{S}$ that use the resource r.

The Shapley value utility leads to an exact potential game with potential

Note. See [7, proposition 2].

$$\phi(\gamma) := \sum_{r \in \mathcal{R}} \sum_{\mathcal{S} \subset \mathcal{S}_r(\gamma)} \frac{1}{|\mathcal{S}|} \left(\sum_{\mathcal{T} \subset \mathcal{S}} (-1)^{|\mathcal{S}| - |\mathcal{T}|} W_r(\mathcal{T}) \right),$$

that typically does not match $W(\gamma)$. In this case, there is no guarantee that the minimum of the global welfare cost is a Nash equilibrium.

Note 11. The game obtained from the Shapley value utility has the advantage (over the one obtained from the wonderful life utility) that it is *budget balanced*. ▷ p. 152

Note 11 (Budget balanced distributed welfare games). The game obtained from the Shapley value utility has the advantage (over the one obtained from the wonderful life utility) that it is *budget balanced* in the sense that the outcomes of the players are of the form

$$J_i(\gamma) = \sum_{r \in \gamma_i} f_r(i, \mathcal{S}_r(\gamma))$$

with the functions $f_r(\cdot)$ (called *distribution rules*) satisfying

$$\sum_{i \in \mathcal{S}} f_r(i, \mathcal{S}) = W_r(\mathcal{S}) \tag{13.18}$$

for every resource $r \in \mathcal{R}$ and every set of players $\mathcal{S} \subset \{1, 2, \ldots, N\}$. Budget balance is important when one needs to match the welfare cost $W_r(\mathcal{S})$ for resource r with the sum of the distribution rules for the players in $\mathcal{S}$. This arises, e.g., when $f_r(i, \mathcal{S})$ is a price (or revenue) that player P_i must pay (or receive) to use the resource r, while this resource is being shared by all the players in $\mathcal{S}$. In this case, the budget balance in (13.18) means that the total payments (or revenues) for each resource r match precisely the welfare cost (or revenue) of that resource.

While budget balanced, the Shapley value utility has the difficulty that its evaluation is computationally difficult for a large number of players since the summation over all subsets $\mathcal{S}$ of $\mathcal{S}_r(\gamma) \setminus \{i\}$ may include a very large number of terms when the number of players is large. □

13.6.2 Price of Anarchy

For either the wonderful life utility or for the Shapley value utility, there is no guarantee that noncooperative players will select policies that minimize the global welfare cost. With the wonderful life utility this is because there may be directionally-local minima that are not the global minima; and with the Shapley value utility the global minimum may not even be a Nash equilibrium. This observation motivates the following definition:

$$PoA := \frac{\max_{\gamma \in \Gamma_{\text{eq}}} W(\gamma)}{\min_{\gamma \in \Gamma} W(\gamma)}, \tag{13.19}$$

where Γ_{eq} denotes the set of Nash equilibria of the distributed welfare game. The ratio in (13.19) is known as the *price of anarchy* and compares the value of the global welfare cost for the "worst" Nash equilibrium (i.e., the one corresponding to the largest cost) with the global minimum of the global welfare cost. A large price of anarchy means that when the players engage in the noncooperative game, they may end up at a Nash equilibrium that is significantly larger than the global minimum—this can be viewed as the price to pay for behaving as cooperative agents.

It turns out that for both the wonderful life utility and for the Shapley value utility, the price of anarchy cannot exceed 2 under mild assumptions on the welfare costs:

Note 12. A cost function defined on sets is *supermodular* if, as a set enlarges, the marginal cost associated with adding an element to the set increases. ▹ p. 153

Lemma 13.1. Consider a distributed welfare game with outcomes given by (13.15) or (13.17) and assume that the welfare cost W_r of every resource $r \in \mathcal{R}$ is supermodular, then $PoA \leq 2$. □

Note 12 (Supermodularity). A cost function W that maps subsets of $\mathcal{D}$ to real numbers is called *supermodular* if for every $i \in \mathcal{D}$ and subsets $\mathcal{S}_1, \mathcal{S}_2 \subset \mathcal{D}$, we have that

$$\mathcal{S}_1 \subset \mathcal{S}_2 \Rightarrow W(\mathcal{S}_1 \cup \{i\}) - W(\mathcal{S}_1) \leq W(\mathcal{S}_2 \cup \{i\}) - W(\mathcal{S}_2).$$

Supermodularity means that, as a set enlarges, the marginal cost associated with adding an element to the set increases (or at least does not decrease). □

13.7 Computation of Nash Equilibria for Potential Games

Note. Nash equilibria always exist, e.g., whenever the action spaces are finite (see Proposition 12.2). ▹ p. 135

An attractive feature of potential games is that, not only do they typically have at least one Nash equilibrium, but also it is straightforward to construct algorithms to compute such equilibria. Next we discuss one such algorithm that is applicable to any potential game with finite action spaces.

Consider a game with N players $\mathsf{P}_1, \mathsf{P}_2, \dots, \mathsf{P}_N$, which select policies from the *action spaces* $\Gamma_1, \Gamma_2, \dots, \Gamma_N$, respectively. As usual, when each P_i uses a policy $\gamma_i \in \Gamma_i$, we denote by

$$J_i(\gamma), \qquad \gamma := (\gamma_1, \gamma_2, \dots, \gamma_N) \in \Gamma := \Gamma_1 \times \Gamma_2 \times \cdots \Gamma_N$$

the *outcome of the game* for the player P_i, and all players want to minimize their own outcomes.

A *(pure) path* is a sequence of multiplayer policies

$$\{\gamma^1, \gamma^2, \dots\}, \qquad \gamma^k = (\gamma_1^k, \gamma_2^k, \dots, \gamma_N^k) \in \Gamma := \Gamma_1 \times \Gamma_2 \times \cdots \times \Gamma_N$$

where each $\gamma^{k+1} \in \Gamma$ differs from the previous $\gamma^k \in \Gamma$ by change in the policy of a single player P_i, i.e., for every k, there exists a player P_i, $i \in \{1, 2, \dots, N\}$ for which

$$\gamma^{k+1} = (\gamma_i, \gamma_{-i}^k), \qquad \gamma_i \neq \gamma_i^k.$$

A path is said to be an *improvement path* with respect to a game with outcomes $J_i(\gamma)$ if, for every k,

$$\gamma^{k+1} = (\gamma_i, \gamma_{-i}^k), \ \gamma_i \neq \gamma_i^k \ \Rightarrow \ J_i(\gamma^{k+1}) < J_i(\gamma^k),$$

which means that if γ^{k+1} differs from γ^k due to a change in policy of player P_i, then this change in policy must result in a strict improvement of player P_i's outcome. Improvement paths can only terminate when no player can improve their outcome by changing policy.

Generating improvement paths is straightforward and can be accomplished by the following algorithm:

MATLAB® Hint 4. Improvement paths can be constructed easily in MATLAB® for general N-player games. ▷ p. 155

1. Pick an arbitrary initial multiplayer policy γ^1 and set $k = 1$.
2. Find a player P_i for which there exists a policy $\gamma_i \in \Gamma_i$ such that

$$J_i(\gamma_i, \gamma_{-i}^k) < J_i(\gamma^k).$$

3. If such a player does not exist, terminate the path.
4. Otherwise, set

$$\gamma^{k+1} = (\gamma_i^{k+1}, \gamma_{-i}^k), \gamma_i^{k+1} \in \arg\min_{\gamma_i \in \Gamma_i} J_i(\gamma_i, \gamma_{-i}^k),$$

increase k by 1 and repeat from step 2.

Note. The policy γ_i^{k+1} in step 4 can be viewed as P_i's best response against γ_{-i}^k. When the minimum is achieved at multiple policies γ_i, one can pick γ_i^{k+1} to be any of them.

For finite exact or ordinal potential games, improvement paths always terminate and lead to a Nash equilibrium:

Proposition 13.1 (Finite improvement property). Every finite improvement path terminates at a Nash equilibrium. Moreover, every improvement path of an exact or an ordinal potential game with finite action spaces is finite and therefore terminates at a Nash equilibrium. □

Proof of Proposition 13.1. The fact that a finite improvement path terminates at a Nash equilibrium is a consequence of the fact that improvement paths are only allowed to terminate when no player can unilaterally improve their own outcome, which means that if an improvement path terminates at a step k, then we must have that

$$J_i(\gamma_i^k, \gamma_{-i}^k) \leq J_i(\gamma_i, \gamma_{-i}^k), \quad \forall \gamma_i \in \Gamma_i, \ \ i \in \{1, 2, \dots, N\},$$

since otherwise we could continue the path. This shows that $(\gamma_i^k, \gamma_{-i}^k)$ is a Nash equilibrium.

To show that every improvement path must terminate, note that for every γ^k along an improvement path, there exists a player P_i such that

$$J_i(\gamma^{k+1}) < J_i(\gamma^k),$$

and therefore, either for an exact or an ordinal potential game with potential ϕ, we must have

$$\phi(\gamma^{k+1}) < \phi(\gamma^k),$$

which means that the potential must be strictly decreasing along any improvement path. When the action spaces are finite, the potential function can only take a finite number of values and therefore can only decrease strictly a finite number of times. This means a finite improvement path must always be finite. ■

MATLAB® Hint 4 (Improvement path). The algorithm outlined to construct improvement paths can be implemented easily in MATLAB®. To this effect, suppose that we enumerate the actions of all players so that every action space can be represented by a numerical array and that the following variables have been constructed within MATLAB®:

- `actionSpaces` is a cell array with N entries, each containing the action space for one player. Specifically, `actionSpaces{i}`, $i \in \{1, 2, \ldots, N\}$ is a numerical array with all the actions available to player P_i.
- `A0` is a matrix with N entries containing the actions of the players at the start of the path. Specifically, `A0(i)`, $i \in \{1, 2, \ldots, N\}$ is the action of P_i at the start of the path.
- `funJ` is a MATLAB® function that returns an N-element matrix with the players' costs for a given matrix of players' actions. Specifically, given an N-element matrix `A` with the actions of all the players, the costs' matrix is obtained using `funJ(A)`.

With these definitions, the following MATLAB® function, returns an improvement path. It uses randomization to make sure that, if it is called multiple times, it will likely result in distinct improvement paths and potentially distinct Nash equilibria.

```
function [A,J]=improvementPath(actionSpaces,A0)
    N=prod(size(actionSpaces)); % number of players
    J=funJ(A0); % compute the initial costs for every player
    A=A0;
    improved=true;
    while improved
        improved=false;
        for i=randperm(N) % range over players in random order
            actionSpace=setdiff(actionSpaces{i},A(i));
            % range over alternative actions for player i in random order
            for newAi=actionSpace(randperm(length(actionSpace)))
                oldAi=A(i);  % save previous action
                A(i)=newAi;  % try alternative action
```

```
                    newJ=funJ(A);
                    if newJ(i)<J(i)
                        % new action improves outcome for player i, keep it
                        J=newJ;
                        improved=true;
                    else
                        A(i)=oldAi; % new action does not improve, discard it
                    end
                end % for newAi=...
                if improved
                    break; % found improvement: try all players again
                end
            end % for i=...
        end % while improved
end
```

This function returns the following variables:

- A is a matrix with N entries with the players' actions at the end of the improvement path. In view of Proposition 13.1, this must correspond to a Nash equilibrium.
- J is a matrix with N entries with the players' costs at the end of the improvement path. □

13.8 Fictitious Play

Consider a game with N players $\mathsf{P}_1, \mathsf{P}_2, \ldots, \mathsf{P}_N$ that select probability distributions from mixed action spaces $\mathcal{Y}^1, \mathcal{Y}^2, \ldots, \mathcal{Y}^N$, respectively. As usual, when each P_i uses a mixed policy $y_i \in \mathcal{Y}^i$, we denote by

$$J_i(y), \qquad y := (y_1, y_2, \ldots, y_N) \in \Gamma := \mathcal{Y}^1 \times \mathcal{Y}^2 \times \cdots \mathcal{Y}^N$$

the *outcome of the game* for the player P_i and all players want to minimize their own outcomes.

A *mixed path* is a sequence of multiplayer policies

$$\{y^1, y^2, \ldots\}, \qquad y^k = (y_1^k, y_2^k, \ldots, y_N^k) \in \mathcal{Y} := \mathcal{Y}^1 \times \mathcal{Y}^2 \times \cdots \times \mathcal{Y}^N$$

that the players use over consecutive repetitions of the game.

In *fictitious play*, all the players construct "beliefs" regarding the mixed policies used by the remaining players as the path $\{y^1, y^2, \ldots\}$ progresses. To accomplish this, they keep track of the empirical distributions of the mixed policies used by the other players up to the current time. Specifically, the *belief* at time k regarding the mixed policy used by player P_i is defined by the time-average of the mixed policies used by P_i up to time k:

$$\hat{y}_i^k := \frac{1}{k} \sum_{\ell=1}^{k} y_i^\ell, \qquad \forall k \geq 1.$$

In fictitious play, all the players then assume that these beliefs are correct, in the sense that the remaining players will use a mixed policy that is exactly equal to these beliefs, and then every player selects a mixed policy that is a "best response" to these beliefs. Specifically, player P_i selects a mixed policy y_i^{k+1} for the next time step that satisfies

$$J_i(y_i^{k+1}, \hat{y}_{-i}^k) \le J_i(y_i, \hat{y}_{-i}^k), \quad \forall y_i \in \mathcal{Y}^i,$$

or equivalently, y_i^{k+1} is the best response against $\hat{y}_{-i}^k$:

$$y_i^{k+1} \in \arg\min_{y_i \in \mathcal{Y}^i} J_i(y_i, \hat{y}_{-i}^k).$$

MATLAB® Hint 5. Fictitious play can be implemented easily in MATLAB® for general N-player games. ▷ p. 158

Clearly the assumption that the beliefs precisely match the mixed policies that the other players will use is not correct under fictitious play, because all players keep adjusting their own mixed policies. Nevertheless, for potential games, fictitious play leads to beliefs $\hat{y}^k$ that converge to Nash equilibria:

Note. See [8, theorem A] and [9, theorem 2.4].

Note. When the Nash equilibrium is unique, Theorem 13.1 simplifies to the statement that the belief sequence converges to the Nash equilibrium.

Theorem 13.1 (Fictitious play for potential games). For every (exact) potential game with mixed policies, the fictitious play belief sequence converges to the set of mixed Nash equilibria for the game, in the sense that each limit point of the belief sequence $\hat{y}^k$

$$\{\hat{y}^1, \hat{y}^2, \ldots\}, \qquad \hat{y}^k = (\hat{y}_1^k, \hat{y}_2^k, \ldots, \hat{y}_N^k) \in \mathcal{Y} := \mathcal{Y}^1 \times \mathcal{Y}^2 \times \cdots \times \mathcal{Y}^N$$

is a Nash equilibrium. □

It turns out that the use of fictitious play goes much beyond potential games, as one can conclude from the following two results:

Theorem 13.2 (Fictitious play for zero-sum games [12]). For every finite zero-sum game, the fictitious play belief sequence

$$\{\hat{y}^1, \hat{y}^2, \ldots\}, \qquad \hat{y}^k = (\hat{y}_1^k, \hat{y}_2^k, \ldots, \hat{y}_N^k) \in \mathcal{Y} := \mathcal{Y}^1 \times \mathcal{Y}^2 \times \cdots \times \mathcal{Y}^N$$

always converges to a saddle-point equilibrium. □

Theorem 13.3 (Fictitious play for general games [3]). For any finite game, if the belief sequence

$$\{\hat{y}^1, \hat{y}^2, \ldots\}, \qquad \hat{y}^k = (\hat{y}_1^k, \hat{y}_2^k, \ldots, \hat{y}_N^k) \in \mathcal{Y} := \mathcal{Y}^1 \times \mathcal{Y}^2 \times \cdots \times \mathcal{Y}^N$$

converges, that limit of the belief sequence is a Nash equilibrium for the game. □

Note. This well known example is due to Shapley [14].

While fictitious play converges for many games, it does not necessarily converge for every game, including the following bimatrix game

$$A = \begin{bmatrix} 0 & 1 & 0 \\ 0 & 0 & 1 \\ 1 & 0 & 0 \end{bmatrix}, \qquad B = \begin{bmatrix} 0 & 0 & 1 \\ 1 & 0 & 0 \\ 0 & 1 & 0 \end{bmatrix},$$

for which fictitious play does not converge, e.g., starting from the following initial beliefs

$$\hat{y}_1^1 = \begin{bmatrix} 1 \\ 0 \\ 0 \end{bmatrix}, \qquad \hat{y}_2^1 = \begin{bmatrix} 0 \\ 1 \\ 0 \end{bmatrix}.$$

MATLAB® Hint 5 (Fictitious play). Fictitious play can be implemented easily in MATLAB® for general N-player games. To this effect suppose that we constructed an $(N+1)$-dimensional tensor `A` that defines the pure game outcomes. Specifically, the entry `A(i1,i2, . . . ,iN,i)` contains the outcome for player P_i, when

- player P_1 selects the pure policy `i1`,
- player P_2 selects the pure policy `i2`, ...
- player P_N selects the pure policy `iN`.

The following MATLAB® function plays `K` rounds of fictitious play for the game described by the tensor `A`.

```
function [belief,value]=fictitiousPlay(A,K)
    N=length(size(A))-1;
    %% compute random initialization for the mixed path
    belief=cell(N,1);
    for i=1:N
        if 1
            belief{i}=rand(size(A,i),1); % random belief;
        else
            belief{i}=ones(size(A,i),1); % uniform belief;
        end
        belief{i}=belief{i}/sum(belief{i}); % normalize to get distribution
    end
    %% iterate fictitious play
    for k=1:K
        %% compute probability-weighted outcomes
        Ay=A;
        for i=1:N
            reps=size(A);
            reps(i)=1;
            shape=ones(1,N+1);
            shape(i)=length(belief{i});
            yi=reshape(belief{i},shape);
            yi=repmat(yi,reps);
            % for player i's outcomes do not multiply by its own
            % belief to eventually compute the best response
            str=['yi(',repmat(':,',1,N),num2str(i),')=1;'];
            eval(str);
            Ay=Ay.*yi;
        end % for i=1:N
        %% compute best response
        y=cell(N,1);
        for i=1:N
            % average outcomes over actions of other players
```

```
            str=['Ay(',repmat(':,',1,N),num2str(i),')'];
            S=eval(str);
            for j=1:N
                if j~=i
                    S=sum(S,j);
                end
            end
            % compute best responses
            [~,j]=min(S,[],i); % pure best response
            y{i}=zeros(size(S,i),1);
            y{i}(j)=1;                  % mixed best response
        end % for i=1:N
        %% update belief
        for i=1:N
            belief{i}=(k*belief{i}+y{i})/(k+1);
        end % for i=1:N
    end  % for k=1:K
    %% compute final value
    value=nan(N,1);
    % compute probability-weighted outcomes
    Ay=A;
    for i=1:N
        reps=size(A);
        reps(i)=1;
        shape=ones(1,N+1);
        shape(i)=length(belief{i});
        yi=reshape(belief{i},shape);
        yi=repmat(yi,reps);
        Ay=Ay.*yi;
    end % for i=1:N
    for i=1:N
        % average outcomes over actions of all players
        str=['Ay(',repmat(':,',1,N),num2str(i),')'];
        S=eval(str);
        for j=1:N
            S=sum(S,j);
        end
        value(i)=S;
    end % for i=1:N
end
```

This function returns an N-dimensional cell array `belief` with the players' beliefs at the end of the path. Specifically, the entry *belief{i}* is the belief for player P_i, which is a probability distribution over the probability simplex of dimension equal to `size(A,i)`. It also returns a vector with the value of the game for the beliefs at the end of the path. □

13.9 Practice Exercises

13.1. Verify that (13.3) is an exact potential for the game with outcomes given by (13.2).

Solution to Exercise 13.1. For every $\gamma_i, \bar{\gamma}_i \in \Gamma_i, \gamma_{-i} \in \Gamma_{-i}$, and $i \in \{1, 2, \dots, N\}$, we have that

$$J_i(\gamma_i, \gamma_{-i}) - J_i(\bar{\gamma}_i, \gamma_{-i}) = H_i(\gamma_i) - H_i(\bar{\gamma}_i) = \phi(\gamma_i, \gamma_{-i}) - \phi(\bar{\gamma}_i, \gamma_{-i})$$

which confirms that ϕ is an exact potential function for the game.

13.2 (Sudoku). Consider the multi-player version of the Sudoku game discussed in Section 13.5 with the outcome of player P_i given by

$$J_i(\gamma) := \sigma_i^{\text{row}}(\gamma) + \sigma_i^{\text{col}}(\gamma) + \sigma_i^{\text{block}}(\gamma), \qquad \gamma := \{\gamma_1, \gamma_2, \dots, \gamma_N\},$$

where $\sigma_i^{\text{row}}(\gamma)$, $\sigma_i^{\text{col}}(\gamma)$, and $\sigma_i^{\text{block}}(\gamma)$ denote the number of times that the digit γ_i selected by P_i appears elsewhere in this player's row, column, and block, respectively.

Show that this defines an exact potential game with potential

$$\phi(\gamma) := \frac{1}{2}\sum_{j=1}^{N} J_j(\gamma).$$

Solution to Exercise 13.2. We can view this game as the sum of three games: one with outcomes given by $\sigma_i^{\text{row}}(\gamma)$, which we call the *rows game*; another with outcomes given by $\sigma_i^{\text{col}}(\gamma)$, which we call the *columns game*; and finally a third one with outcomes given by $\sigma_i^{\text{block}}(\gamma)$, which we call the *blocks game*. We will show that each of these games is an exact potential game with potentials given by

$$\phi^{\text{row}}(\gamma) := \frac{1}{2}\sum_{j=1}^{N} \sigma_j^{\text{row}}(\gamma), \qquad \phi^{\text{col}}(\gamma) := \frac{1}{2}\sum_{j=1}^{N} \sigma_j^{\text{col}}(\gamma),$$

$$\phi^{\text{block}}(\gamma) := \frac{1}{2}\sum_{j=1}^{N} \sigma_j^{\text{block}}(\gamma),$$

respectively. Once we prove this, it follows from Proposition 12.4 that

$$\phi(\gamma) := \phi^{\text{row}}(\gamma) + \phi^{\text{col}}(\gamma) + \phi^{\text{block}}(\gamma) = \frac{1}{2}\sum_{j=1}^{N} J_j(\gamma)$$

is indeed a potential for the sum game.

To show that the rows game is an exact potential game with potential ϕ^{row}, we pick $\gamma_i \neq \bar{\gamma}_i \in \Gamma_i, \gamma_{-i} \in \Gamma_{-i}, i \in \{1, 2, \dots, N\}$ and compute

$$\begin{aligned}\phi^{\text{row}}(\gamma_i, \gamma_{-i}) - \phi^{\text{row}}(\bar{\gamma}_i, \gamma_{-i}) &= \frac{1}{2}\sum_{j=1}^{N}\left(\sigma_j^{\text{row}}(\gamma_i, \gamma_{-i}) - \sigma_j^{\text{row}}(\bar{\gamma}_i, \gamma_{-i})\right)\\ &= \frac{1}{2}\left(\sigma_i^{\text{row}}(\gamma_i, \gamma_{-i}) - \sigma_i^{\text{row}}(\bar{\gamma}_i, \gamma_{-i})\right)\\ &\quad + \frac{1}{2}\sum_{j\neq i}\left(\sigma_j^{\text{row}}(\gamma_i, \gamma_{-i}) - \sigma_j^{\text{row}}(\bar{\gamma}_i, \gamma_{-i})\right).\end{aligned} \tag{13.20}$$

The only terms in the right-hand side summation for which $\sigma_j^{\text{row}}(\gamma_i, \gamma_{-i}) -\sigma_j^{\text{row}}(\bar{\gamma}_i, \gamma_{-i}) \neq 0$, must correspond to players P_j, $j \neq i$ that are in the same row as player P_i, since for all others the value of σ_j^{row} is not affected by a change in the digit selected by P_i. Among those players P_j, $j \neq i$, we have

$$\sigma_j^{\text{row}}(\gamma_i, \gamma_{-i}) - \sigma_j^{\text{row}}(\bar{\gamma}_i, \gamma_{-i}) = \begin{cases} +1 & \mathsf{P}_j \text{ selected a digit } \gamma_j = \gamma_i \neq \bar{\gamma}_i \\ -1 & \mathsf{P}_j \text{ selected a digit } \gamma_j = \bar{\gamma}_i \neq \gamma_i \\ 0 & \mathsf{P}_j \text{ neither selected } \gamma_j \notin \{\gamma_i, \bar{\gamma}_i\}. \end{cases}$$

Therefore the summation

$$\sum_{j\neq i}\left(\sigma_j^{\text{row}}(\gamma_i, \gamma_{-i}) - \sigma_j^{\text{row}}(\bar{\gamma}_i, \gamma_{-i})\right)$$

is simply equal to the number of players P_j, $j \neq i$ in the same row as P_i that selected γ_i minus the number of players that selected $\bar{\gamma}_i$, which means that

$$\sum_{j\neq i}\left(\sigma_j^{\text{row}}(\gamma_i, \gamma_{-i}) - \sigma_j^{\text{row}}(\bar{\gamma}_i, \gamma_{-i})\right) = \sigma_i^{\text{row}}(\gamma_i, \gamma_{-i}) - \sigma_i^{\text{row}}(\bar{\gamma}_i, \gamma_{-i}).$$

This shows that the two terms in the right-hand side of (13.20) are equal to each other and therefore

$$\phi^{\text{row}}(\gamma_i, \gamma_{-i}) - \phi^{\text{row}}(\bar{\gamma}_i, \gamma_{-i}) = \sigma_i^{\text{row}}(\gamma_i, \gamma_{-i}) - \sigma_i^{\text{row}}(\bar{\gamma}_i, \gamma_{-i}),$$

which establishes that ϕ^{row} is an exact potential for the rows game. The same argument can be used to conclude that ϕ^{col} and ϕ^{block} are exact potentials for the column and block games, respectively, by looking at columns and blocks instead of rows.

13.3. Verify that (13.11) is an exact potential function for the game with outcomes given by (13.9).

Solution to Exercise 13.3. For every $\gamma_i, \bar{\gamma}_i \in \Gamma_i$, $\gamma_{-i} \in \Gamma_{-i}$, and $i \in \{1, 2, \ldots, N\}$, the function ϕ in (13.11) can be re-written as

$$\begin{aligned}
&\phi(\gamma_i, \gamma_{-i}) \\
&\quad = \sum_{k=1}^{N} \left(H_k(\gamma_k) + \sum_{j=1}^{k-1} W_{kj}(\gamma_k, \gamma_j) \right) \\
&\quad = \sum_{k=1}^{i-1} \left(H_k(\gamma_k) + \sum_{j=1}^{k-1} W_{kj}(\gamma_k, \gamma_j) \right) + \left(H_i(\gamma_i) + \sum_{j=1}^{i-1} W_{ij}(\gamma_i, \gamma_j) \right) \\
&\qquad + \sum_{k=i+1}^{N} \left(H_k(\gamma_k) + \sum_{j=1}^{i-1} W_{kj}(\gamma_k, \gamma_j) + W_{ki}(\gamma_k, \gamma_i) + \sum_{j=i+1}^{k-1} W_{kj}(\gamma_k, \gamma_j) \right) \\
&\quad = H_i(\gamma_i) + \sum_{j=1}^{i-1} W_{ij}(\gamma_i, \gamma_j) + \sum_{k=i+1}^{N} W_{ki}(\gamma_k, \gamma_i) + Q_i(\gamma_{-i})
\end{aligned}$$

where $Q_i(\gamma_{-i})$ stands for all the terms that do not depend on γ_i. Using (13.10), we further conclude that

$$\begin{aligned}
\phi(\gamma_i, \gamma_{-i}) &= H_i(\gamma_i) + \sum_{j=1}^{i-1} W_{ij}(\gamma_i, \gamma_j) + \sum_{k=i+1}^{N} W_{ik}(\gamma_i, \gamma_k) + Q_i(\gamma_{-i}) \\
&= H_i(\gamma_i) + \sum_{j \neq i} W_{ij}(\gamma_i, \gamma_j) + Q_i(\gamma_{-i}) = J_i(\gamma_i, \gamma_{-i}) + Q_i(\gamma_{-i})
\end{aligned}$$

and therefore

$$J_i(\gamma_i, \gamma_{-i}) = \phi(\gamma_i, \gamma_{-i}) - Q_i(\gamma_{-i}),$$

which shows that G is the sum of an identical interests game and a dummy game.

13.4. Verify that the game with outcomes (13.15) can be viewed as the sum of an identical interests game with outcome (13.16) with a dummy game, as in (13.1) and Proposition 12.5.

Solution to Exercise 13.4. Since

$$r \notin \gamma_i \;\Rightarrow\; i \notin \mathcal{S}_r(\gamma) \;\Rightarrow\; W_r(\mathcal{S}_r(\gamma) \setminus \{i\}) = W_r(\mathcal{S}_r(\gamma))$$

we can also write

$$J_i(\gamma) = \sum_{r \in \mathcal{R}} \big(W_r(\mathcal{S}_r(\gamma)) - W_r(\mathcal{S}_r(\gamma) \setminus \{i\}) \big) = W(\gamma) - Q_i(\gamma),$$

5	3			7				
6			1	9	5			
	9	8					6	
8		9		6				3
4			8	5	3			1
7			9	2			5	6
	6					2	8	
			4	1	9		3	5
				8			7	9

Figure 13.2 Sudoku puzzle for Exercises 13.5 and 13.9.

where

$$Q_i(\gamma) := \sum_{r \in \mathcal{R}} W_r(\mathcal{S}_r(\gamma) \setminus \{i\}).$$

But $Q_i(\gamma)$ does not depend on γ_i and therefore the game with outcomes (13.15) can be viewed as the sum of an identical interests game with outcomes given by $\phi(\gamma) := W(\gamma)$ and a dummy game with outcomes $Q_i(\gamma)$.

13.5 (Sudoku). Write a MATLAB® script to solve the Sudoku puzzle in Figure 13.2 by computing improvement paths for the multi-player version of the Sudoku game considered in Section 13.5.

Hint: Make use of the code in MATLAB® Hint 4.

Solution to Exercise 13.5. To use the code in MATLAB® Hint 4, we need to construct a 9×9 cell array with the action spaces for all the players. To accomplish this, we start by encoding the Sudoku board layout into a 9×9 matrix with all the digits in the layout and zeros where the layout has an empty slot. For the puzzle in Figure 13.2, this matrix is:

```
board=[5 3 0  0 7 0  0 0 0;
       6 0 0  1 9 5  0 0 0;
       0 9 8  0 0 0  0 6 0;

       8 0 9  0 6 0  0 0 3;
       4 0 0  8 5 3  0 0 1;
       7 0 0  9 2 0  0 5 6;

       0 6 0  0 0 0  2 8 0;
       0 0 0  4 1 9  0 3 5;
       0 0 0  0 8 0  0 7 9];
```

The following MATLAB® function can then be used to construct the cell array with the action spaces:

- whenever a digit appears in the board matrix, the corresponding player has a single action which is precisely that digit; and

- when a 0 appears in the board matrix, the corresponding player can play any digit that does not appear in its row, column, and block.

```
function actionSpaces=sudoku_actionspaces(board);
    actionSpaces=cell(size(board));
    for row=1:9
        for col=1:9
            if board(row,col) ~= 0
                actionSpaces{row,col}=board(row,col);
            else
                actionSpaces{row,col}=1:9;
                % remove digits in same row
                actionSpaces{row,col}=setdiff(actionSpaces{row,col},...
                                              board(row,:));
                % remove digits in same column
                actionSpaces{row,col}=setdiff(actionSpaces{row,col},...
                                              board(:,col));
                % remove digits in same block
                rowBlk=3*floor((row-1)/3)+(1:3);
                colBlk=3*floor((col-1)/3)+(1:3);
                actionSpaces{row,col}=setdiff(actionSpaces{row,col},...
                                                board(rowBlk,colBlk));
            end
        end
    end
end
```

We also need a function funJ that, given a 9×9 matrix A with the players' actions, returns a 9×9 matrix with the corresponding players' costs:

```
function J=funJ(A)
    J=zeros(size(A));
    for row=1:9
        for col=1:9
            rowBlk=3*floor((row-1)/3)+(1:3);
            colBlk=3*floor((col-1)/3)+(1:3);
            J(row,col)=sum(A(row,:)==A(row,col))+...
                       sum(A(:,col)==A(row,col))+...
                       sum(sum(A(rowBlk,colBlk)==A(row,col)))-3;
        end
    end
end
```

Finally, the following MATLAB® code calls the function improvementPath defined in MATLAB® Hint 4 multiple times to try to obtain a Nash equilibrium that is a global minimum of the potential.

```
actionSpaces=sudoku_actionspaces(board);
nRepeats=1000;
for k=1:nRepeats
    % compute random initial actions
    A0=nan(size(actionSpaces));
    for i=1:N
```

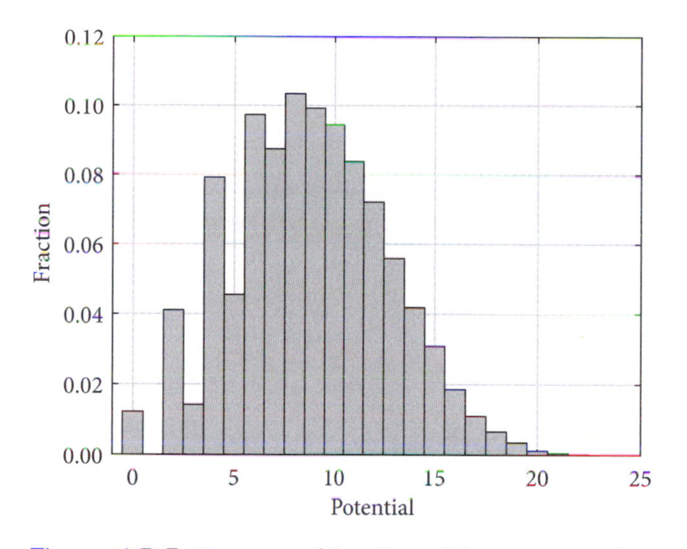

Figure 13.3 Histogram of the values of the potential at the Nash equilibria obtained by computing 10,000 improvement paths for the multi-player version of the Sudoku game in Exercise 13.5.

```
        A0(i)=actionSpaces{i}(randi(length(actionSpaces{i})));
    end
    [A,J]=improvementPath(actionSpaces,A0);
    if sum(J(:))==0
        fprintf('Solution found:\n');
        disp(A);
        break;
    end
end
```

Figure 13.3 shows an histogram of the values of the potential at the Nash equilibria obtained by the code above. We can see that the number of times that we obtain a zero potential—the global minimum—is relatively small (about 1.22%). This means that generally we must compute many improvement paths until we find a solution to the Sudoku puzzle.

Note. We saw in Exercise 10.1 that the quadratic program solver often fails at finding a Nash equilibrium for this game. However, since the prisoners' dilemma is a potential game (see Exercise 12.1), fictitious play is guaranteed to converge to a Nash equilibrium for this game.

13.6 (Fictitious play). Use the fictitious play algorithm in MATLAB® Hint 5 to compute Nash equilibrium policies for the following games:

1. Prisoners' dilemma with

$$A = \underbrace{\begin{bmatrix} 2 & 30 \\ 0 & 8 \end{bmatrix}}_{\text{P}_2 \text{ choices}} \Big\} \text{P}_1 \text{ choices} \qquad B = \underbrace{\begin{bmatrix} 2 & 0 \\ 30 & 8 \end{bmatrix}}_{\text{P}_2 \text{ choices}} \Big\} \text{P}_1 \text{ choices}.$$

2. Battle of the sexes with

$$A = \underbrace{\begin{bmatrix} -2 & 0 \\ 3 & -1 \end{bmatrix}}_{\text{P}_2 \text{ choices}} \Big\} \text{P}_1 \text{ choices} \qquad B = \underbrace{\begin{bmatrix} -1 & 3 \\ 0 & -2 \end{bmatrix}}_{\text{P}_2 \text{ choices}} \Big\} \text{P}_1 \text{ choices}.$$

3. Rock-paper-scissors with

$$A = \underbrace{\begin{bmatrix} -2 & 0 \\ 3 & -1 \end{bmatrix}}_{\text{P}_2 \text{ choices}} \Big\} \text{P}_1 \text{ choices} \qquad B = \underbrace{\begin{bmatrix} -1 & 3 \\ 0 & -2 \end{bmatrix}}_{\text{P}_2 \text{ choices}} \Big\} \text{P}_1 \text{ choices}.$$

Solution to Exercise 13.6.

1. For the prisoners' dilemma, we use the code

```
A=zeros(2,2,2);
A(:,:,1)=[2,30; 0,8];
A(:,:,2)=[2, 0;30,8];
K=10000;
[belief,value]=fictitiousPlay(A,K);
```

which leads to

```
belief{1}=[0.0000;1.0000];
belief{2}=[0.0001;0.9999];
value    =[8.0005;8.0009];
```

2. For the battle of the sexes, we use the code

```
A=zeros(2,2,2);
A(:,:,1)=[-2,1;0,-1];
A(:,:,2)=[-1,3;2,-2];
K=10000;
[belief,value]=fictitiousPlay(A,K);
```

which leads to

```
belief{1}=[1.0000;0.0000];
belief{2}=[1.0000;0.0000];
value    =[-1.9999;-0.9999];
```

3. For the rock paper scissors, we use the code

```
A=zeros(3,3,2);
A(:,:,1)=[0,+1,-1;-1,0,+1;+1,-1,0];
A(:,:,2)=-A(:,:,1);
K=10000;
[belief,value]=fictitiousPlay(A,K);
```

which leads to

```
belief{1}=[0.3365;0.3353;0.3282];
belief{2}=[0.3317;0.3293;0.3390];
value    =[-2.892e-05;2.892e-05];
```

13.10 Additional Exercises

13.7. Verify that (13.3) is an exact potential for the game with outcomes given by (13.4).

13.8 (Sudoku). Consider the multi-player version of the Sudoku game discussed in Section 13.5, but with the outcome of player P_i given by

$$J_i(\gamma) := \sigma_{r_i}^{\text{row}}(\gamma) + \sigma_{c_i}^{\text{col}}(\gamma) + \sigma_{b_i}^{\text{block}}(\gamma), \qquad \gamma := \{\gamma_1, \gamma_2, \dots, \gamma_N\},$$

where $r_i \in \{1, 2, \dots, 9\}$, $c_i \in \{1, 2, \dots, 9\}$, and $b_i \in \{1, 2, \dots, 9\}$ denote player P_i's row, column, and block, respectively; and $\sigma_r^{\text{row}}(\gamma)$ denotes the total number of times that a digit is repeated in the rth row; $\sigma_c^{\text{col}}(\gamma)$ denotes the number of times that a digit is repeated in the cth column; and $\sigma_b^{\text{block}}(\gamma)$ the number of times that a digit is repeated in the bth block.

Show that this defines an exact potential game potential

$$\phi(\gamma) := \frac{1}{9}\sum_{i=1}^{N} J_i(\gamma) = \sum_{r=1}^{9} \sigma_r^{\text{row}}(\gamma) + \sum_{c=1}^{9} \sigma_c^{\text{col}}(\gamma) + \sum_{b=1}^{9} \sigma_b^{\text{block}}(\gamma).$$

13.9 (Sudoku). Write a MATLAB® script to solve the Sudoku puzzle in Figure 13.2 by computing improvement paths for the multi-player version of the Sudoku game considered in Section 13.5, but with the outcomes in Exercise 13.8.

Hint: Make use of the code in MATLAB®Hint 4.

13.10. Verify that (13.12) is a potential function for the congestion game with outcomes given by (13.13).

PART IV DYNAMIC GAMES

LECTURE 14

Dynamic Games

This lecture introduces a new class of games, called dynamic games, that will be further explored in subsequent lectures.

14.1 Game Dynamics

Consider a two-player multi-stage game in extensive form like the one in Figure 14.1(a) and suppose that we use the following notation: For each stage $k \in \{1, 2, \dots, K\}$,

Notation. One generally calls x_k the *state of a game* at the kth stage.

1. x_k denotes the node at which the game enters the kth stage,
2. u_k denotes the action of player P_1 at the kth stage,
3. d_k denotes the action of player P_2 at the kth stage.

Attention! Equation (14.1) describes the tree itself, but not the outcomes or the information sets.

In this case, the *overall tree structure* can be mathematically described by equations of the form:

$$\underbrace{x_{k+1}}_{\substack{\text{entry node}\\ \text{at stage } k+1}} = \underbrace{f_k}_{\substack{\text{"dynamics" at}\\ \text{stage } k}} \Big(\underbrace{x_k}_{\substack{\text{entry node}\\ \text{at stage } k}}, \underbrace{u_k}_{\substack{\mathsf{P}_1\text{'s action}\\ \text{at stage } k}}, \underbrace{d_k}_{\substack{\mathsf{P}_2\text{'s action}\\ \text{at stage } k}} \Big), \quad \forall k \in \{1, 2, \dots, K-1\}, \tag{14.1}$$

as shown in Table 14.1. This type of description actually allows for games that are more general than those that are typically described in extensive form. For example:

Notation. A tree is a (connected) graph that has no cycles.

- games described by graphs that are not trees, such as the one in Figure 14.1(b);
- games with infinitely many stages ($K = \infty$);
- games with action spaces that are not finite sets.

Notation. Often one regards the stage index k as time and calls these *discrete-time* dynamic games, in contrast with differential games that take place in continuous times, which we will encounter shortly.

Games whose evolution is represented by an equation such as (14.1) are called *dynamic games* and the equation (14.1) is often called the *dynamics of the game*. The set $\mathcal{X}$ where the state x_k takes values is called the *state-space* of the game.

The *outcome* J_i for a particular player P_i, $i \in \{1, 2\}$ in a multi-stage game in extensive form like the one in Figure 14.1(a) is a function of the state of the game

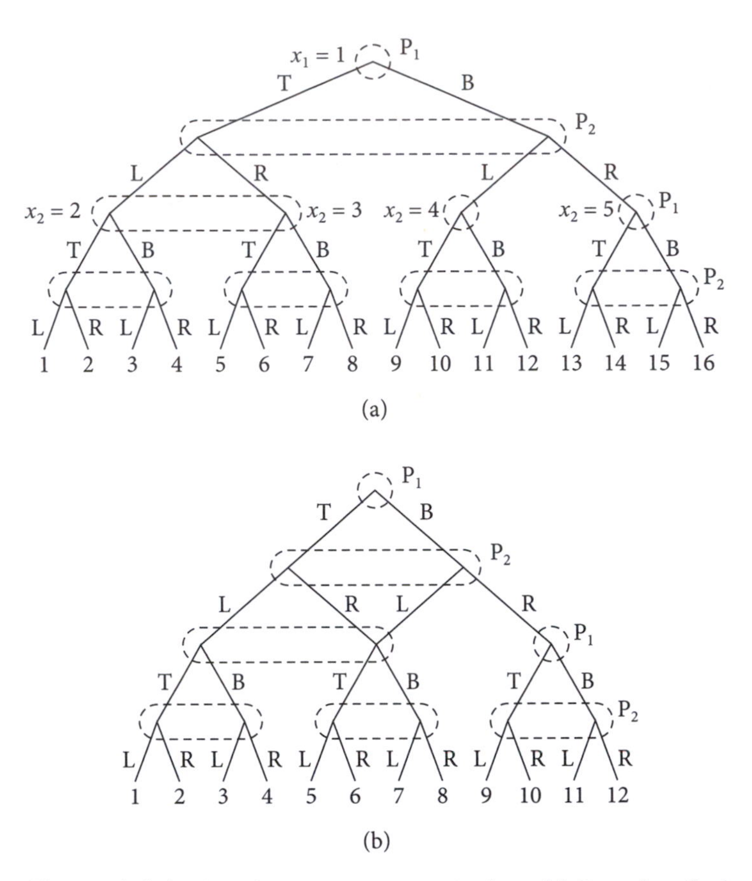

Figure 14.1 Two-player games in extensive form. (a) Game described by a tree; (b) Game described by a graph that is not a tree.

at the last stage K and the actions taken by the players at this stage:

$$J_i(x_K, u_K, d_K), \tag{14.2}$$

as shown in Table 14.1. However, when the game is described by a graph that is not a tree, one may have different outcomes depending on how one got to the end of the game. In this case, the outcome J_i may depend on all the decisions made by both players from the start of the game:

$$J_i(u_1, d_1, u_1, d_1, \cdots, u_K, d_K).$$

A dynamic game is said to have a *stage-additive cost* when the outcome J_i to be *minimized* can be written as

Notation. When $K = \infty$ we have an *infinite horizon game*, in which case (14.3) is really a series.

$$\sum_{k=1}^{K} g_k^i(x_k, u_k, d_k). \tag{14.3}$$

Note. The outcome in (14.2) corresponds precisely to a terminal cost.

When all g_k^i are equal to zero, except for the last g_K^i, the game is said to have a *terminal cost*.

TABLE 14.1 Dynamics and outcome for player P_1 for the 2-stage game in extensive form ($K=2$) shown in Figure 14.1(a).

k	x_k	u_k	d_k	$f_k(x_k, u_k, d_k)$
1	1	T	L	2
1	1	T	R	3
1	1	B	L	4
1	1	B	R	5

x_K	u_K	d_K	$J_1(x_K, u_K, d_K)$
2	T	L	1
2	T	R	2
2	B	L	3
2	B	R	4
3	T	L	5
3	T	R	6
3	B	L	7
3	B	R	8
4	T	L	9
4	T	R	10
4	B	L	11
4	B	R	12
5	T	L	13
5	T	R	14
5	B	L	15
5	B	R	16

14.2 Information Structures

Dynamic games can have a wide range of information structures, but we will consider mostly two structures: open loop and (perfect) state feedback.

In *open-loop (OL)* dynamic games, players do not gain any information as the game is played (other than the current stage) and must make their decisions solely based on a priori information. In terms of an extensive form representation, each player has a single information set per stage, which contains all the nodes for that player at that stage, as in the game shown in Figure 14.2(a). For open-loop dynamic games, one typically represents *policies* as functions of the initial state x_1: when player P_1 uses an open-loop policy $\gamma^{\mathrm{OL}} := \{\gamma_1^{\mathrm{OL}}, \gamma_2^{\mathrm{OL}}, \dots, \gamma_K^{\mathrm{OL}}\}$, that player sets

Notation. By expressing open-loop policies as functions of a (typically fixed) initial state one emphasizes that these policies *cannot* depend on information collected later in the game, in contrast to state-feedback games.

$$u_1 = \gamma_1^{\mathrm{OL}}(x_1), \quad u_2 = \gamma_2^{\mathrm{OL}}(x_1), \quad \cdots \quad u_K = \gamma_K^{\mathrm{OL}}(x_1);$$

and when P_2 uses an open-loop policy $\sigma^{\mathrm{OL}} := \{\sigma_1^{\mathrm{OL}}, \sigma_2^{\mathrm{OL}}, \dots, \sigma_K^{\mathrm{OL}}\}$, that player sets

$$d_1 = \sigma_1^{\mathrm{OL}}(x_1), \quad d_2 = \sigma_2^{\mathrm{OL}}(x_1), \quad \cdots \quad d_K = \sigma_K^{\mathrm{OL}}(x_1).$$

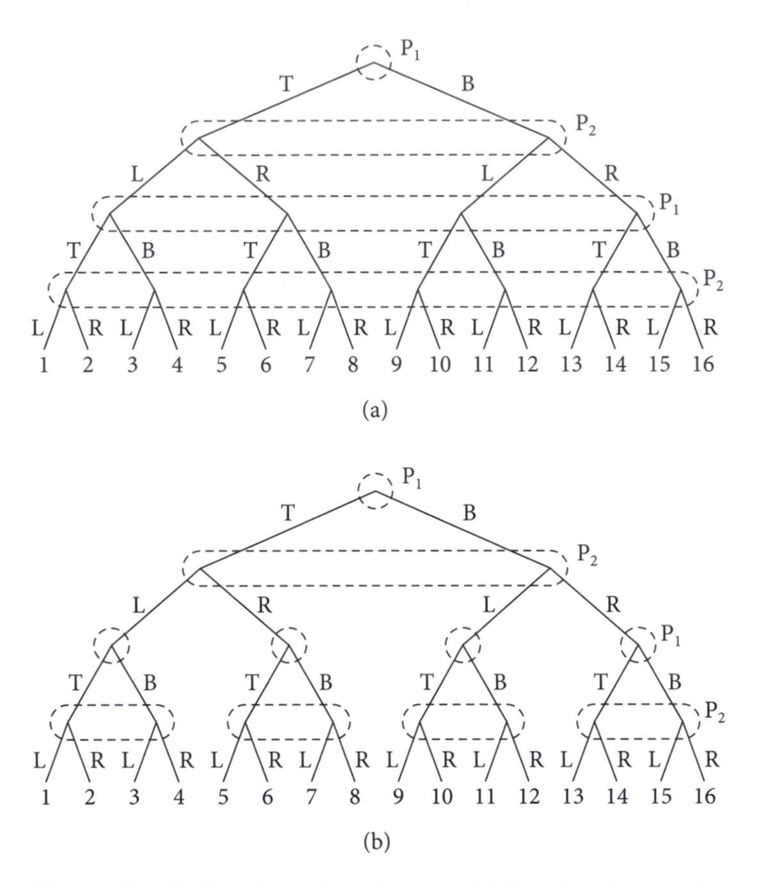

Figure 14.2 Two-player dynamic games. (a) Open-loop information structure; (b) State-feedback information structure.

Notation. The qualifier "perfect" is sometimes used to emphasize that the players know exactly the current state x_k, in contrast to situations in which the players only have access to estimates of x_k, possibly constructed from noisy measurements.

Note. State-feedback games are a special case of the (pure) feedback games that we encountered in Section 7.7, but now the players must select their actions based only on the state of the game, whereas before the information sets could be richer for the player that plays second.

In *(perfect) state-feedback (FB)* games, the players know exactly the state x_k of the game at the entry of the current stage and can use this information to choose their actions u_k and d_k at that stage. However, they must make these decisions without knowing each others choice (i.e., we have simultaneous play at each stage). In terms of an extensive form representation, at each stage of the game there is exactly one information set for each entry-point to that stage, as in the game shown in Figure 14.2(b). For state-feedback games, one typically represents *policies* as functions of the current state: when player P_1 uses a state-feedback policy $\gamma^{\text{FB}} := \{\gamma_1^{\text{FB}}, \gamma_2^{\text{FB}}, \ldots, \gamma_K^{\text{FB}}\}$, that player sets

$$u_1 = \gamma_1^{\text{FB}}(x_1), \quad u_2 = \gamma_2^{\text{FB}}(x_2), \quad \ldots \quad u_K = \gamma_K^{\text{FB}}(x_K);$$

and when P_2 uses a state-feedback policy $\sigma := \{\sigma_1^{\text{FB}}, \sigma_2^{\text{FB}}, \ldots, \sigma_K^{\text{FB}}\}$, that player sets

$$d_1 = \sigma_1^{\text{FB}}(x_1), \quad d_2 = \sigma_2^{\text{FB}}(x_2), \quad \ldots \quad d_K = \sigma_K^{\text{FB}}(x_K).$$

Now that we have defined admissible sets of policies (i.e., action spaces) and how these translate to outcomes through the dynamics of the game, the general definitions introduced in Lecture 9 specify unambiguously what is meant by a security policy or a Nash equilibrium for these games.

14.3 Continuous-Time Differential Games

Dynamic games are often also formulated in continuous time, which means that

1. the state $x(t)$ varies continuously with time on a given interval $t \in [0, T]$, and
2. the players continuously select actions $u(t)$ and $d(t)$ on $[0, T]$, which determine the state's evolution.

When the state $x(t)$ is an n-vector of real numbers whose evolution is determined by a differential equation, the game is called a *differential game*. We consider here differential games with *dynamics* of the form

$$\underbrace{\dot{x}(t)}_{\substack{\text{state}\\ \text{derivative}}} = \underbrace{f}_{\substack{\text{game}\\ \text{dynamics}}}\Big(\underbrace{t}_{\text{time}}, \underbrace{x(t)}_{\substack{\text{current}\\ \text{state}}}, \underbrace{u(t)}_{\substack{\text{P}_1\text{'s action}\\ \text{at time } t}}, \underbrace{d(t)}_{\substack{\text{P}_2\text{'s action}\\ \text{at time } t}}\Big), \quad \forall t \in [0, T], \tag{14.4}$$

Notation. When $T = \infty$ we have an *infinite horizon game* and in this case the final cost term is generally absent.

for which each player P_i, $i \in \{1, 2\}$ wants to *minimize* a *cost* of the form

$$J_i := \underbrace{\int_0^T g_i(t, x(t), u(t), d(t))dt}_{\text{cost along trajectory}} + \underbrace{q_i(x(T))}_{\text{final cost}}. \tag{14.5}$$

For such games we shall also consider *open-loop* policies of the form

$$u(t) = \gamma^{\text{OL}}(t, x(0)), \quad d(t) = \sigma^{\text{OL}}(t, x(0)), \quad \forall t \in [0, T],$$

and *(perfect) state-feedback* policies of the form

$$u(t) = \gamma^{\text{FB}}(t, x(t)), \quad d(t) = \sigma^{\text{FB}}(t, x(t)), \quad \forall t \in [0, T].$$

Example 14.1 (Zebra in the lake). Consider the zebra in the lake game introduced in Section 2.3 and depicted in Figure 14.3, where

1. the player P_1 is a *zebra* that swims with a speed of v_{zebra} in a circular lake with radius R, and
2. the player P_2 is a *lion* that runs along the perimeter of the lake with maximum speed of $v_{\text{lion}} > v_{\text{zebra}}$.

Denoting by $(x_{\text{zebra}}, y_{\text{zebra}})$ the position of the zebra and by θ_{zebra} the orientation, we have that

$$\dot{x}_{\text{zebra}} = v_{\text{zebra}} \cos \theta_{\text{zebra}}, \qquad \dot{y}_{\text{zebra}} = v_{\text{zebra}} \sin \theta_{\text{zebra}}, \qquad \theta_{\text{zebra}} \in [0, 2\pi), \tag{14.6}$$

and denoting by θ_{lion} and ω_{lion} the (angular) position and velocity of the lion, respectively, we have that

$$\dot{\theta}_{\text{lion}} = \omega_{\text{lion}}, \qquad \omega_{\text{lion}} \in \left[-\frac{v_{\text{lion}}}{R}, +\frac{v_{\text{lion}}}{R}\right]. \tag{14.7}$$

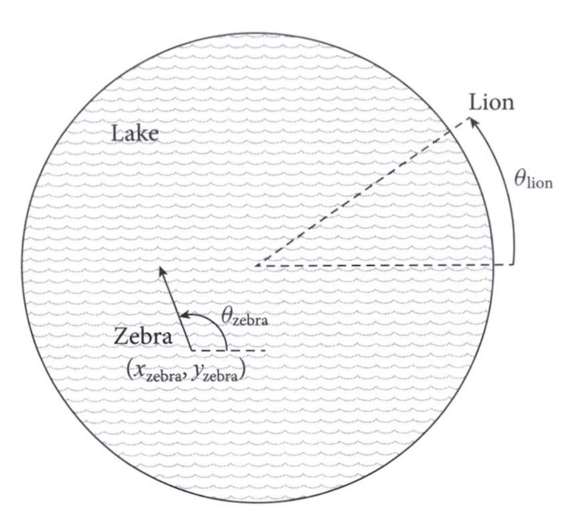

Figure 14.3 The zebra in the lake game.

Defining a state vector

$$x(t) := [\, x_{\text{zebra}}(t) \quad y_{\text{zebra}}(t) \quad \theta_{\text{lion}}(t) \,]',$$

the equations (14.6)–(14.7) can be written as in (14.4), where the actions of the players are:

$$u(t) = \theta_{\text{zebra}}(t) \in [0, \pi), \qquad d(t) = \omega_{\text{lion}}(t) \in \left[-\frac{v_{\text{lion}}}{R}, +\frac{v_{\text{lion}}}{R} \right].$$

If we assume that the zebra wants to get out of the lake as soon as possible without being captured, the zebra's cost is of the form

$$J_1 = \begin{cases} T_{\text{exit}} & \text{zebra exits the lake safely at time } T_{\text{exit}} \\ +\infty & \text{zebra gets caught when it exits.} \end{cases}$$

This is a zero-sum game and therefore the lion wants to maximize J_1, or equivalently minimize $J_2 := -J_1$.

A common trick that is used to write such a cost in an integral form such as (14.5) is to freeze the state when the zebra reaches the shore, which amounts to replacing (14.6)–(14.7) by

Note. There are more compact ways to formalize this game that require a smaller state. For example, because of the radial symmetry, one may use a coordinate system that rotates with the lion and always keeps $\theta_{\text{lion}} = 0$. However, in this case the angular velocity of the lion also affects the equations of motion for the zebra, through a rotation of the coordinate system. One may also use a polar coordinate system to express the position of the zebra [1].

$$\begin{bmatrix} \dot{x}_{\text{zebra}} \\ \dot{y}_{\text{zebra}} \\ \dot{\theta}_{\text{lion}} \end{bmatrix} = \begin{cases} \begin{bmatrix} v_{\text{zebra}} \cos \theta_{\text{zebra}} \\ v_{\text{zebra}} \sin \theta_{\text{zebra}} \\ \omega_{\text{lion}} \end{bmatrix} & x_{\text{zebra}}^2 + y_{\text{zebra}}^2 < R^2 \\ \begin{bmatrix} 0 \\ 0 \\ 0 \end{bmatrix} & x_{\text{zebra}}^2 + y_{\text{zebra}}^2 = R^2, \end{cases}$$

and then defining

$$J_1 := \int_0^\infty g(x_{\text{zebra}}, y_{\text{zebra}}, \theta_{\text{lion}}) dt$$

where

$$g(x_{\text{zebra}}, y_{\text{zebra}}, \theta_{\text{lion}}) = \begin{cases} 1 & x_{\text{zebra}}^2 + y_{\text{zebra}}^2 < R^2 \\ 1 & x_{\text{zebra}} = R\cos\theta_{\text{lion}},\ y_{\text{zebra}} = R\sin\theta_{\text{lion}} \text{ (zebra is caught)} \\ 0 & \text{otherwise (zebra reaches shore away from lion).} \end{cases}$$

This game is only meaningful in the context of state-feedback policies, because the lion basically has no chance of capturing the zebra unless the lion can see the zebra.

□

14.4 Differential Games with Variable Termination Time

A less convoluted way to formalize pursuit-evasion games such as the zebra in the lake is to consider the usual continuous-time dynamics as in (14.4), but costs to be minimized by each player P_i of the form

$$J_i := \underbrace{\int_0^{T_{\text{end}}} g_i(t, x(t), u(t), d(t))dt}_{\text{cost along trajectory}} + \underbrace{q_i(T_{\text{end}}, x(T_{\text{end}}))}_{\text{final cost}}$$

Notation. We can think of $\mathcal{X}_{\text{end}}$ as the set of states at which the game terminates, as the evolution of $x(t)$ is irrelevant after this time. The states in $\mathcal{X}_{\text{end}}$ are often called the *game-over* states.

where T_{end} is the first time at which the state $x(t)$ enters a closed set $\mathcal{X}_{\text{end}} \subset \mathbb{R}^n$ or $T_{\text{end}} = +\infty$ in case $x(t)$ never enters $\mathcal{X}_{\text{end}}$.

Example 14.2 (Zebra in the lake, continued). The zebra in the lake game in Example 14.1 could also be formalized as a differential game with dynamics

$$\begin{bmatrix} \dot{x}_{\text{zebra}} \\ \dot{y}_{\text{zebra}} \\ \dot{\theta}_{\text{lion}} \end{bmatrix} = \begin{bmatrix} v_{\text{zebra}} \cos\theta_{\text{zebra}} \\ v_{\text{zebra}} \sin\theta_{\text{zebra}} \\ \omega_{\text{lion}} \end{bmatrix}, \theta_{\text{zebra}} \in [0, \pi), \omega_{\text{lion}} \in \left[-\frac{v_{\text{lion}}}{R}, +\frac{v_{\text{lion}}}{R}\right]$$

and a cost

$$J_1 := \int_0^{T_{\text{end}}} dt + q(x(T_{\text{end}})),$$

where T_{end} is the first time at which the state $x(t)$ enters the set

$$\mathcal{X}_{\text{end}} := \left\{(x_{\text{zebra}}, y_{\text{zebra}}, \theta_{\text{lion}}) \subset \mathbb{R}^3 : x_{\text{zebra}}^2 + y_{\text{zebra}}^2 \geq R^2\right\}$$

of "safe" configurations for the zebra to reach the shore and the final cost

$$q(x) := \begin{cases} 0 & \text{if } (x_{\text{zebra}}, y_{\text{zebra}}) \neq (R\cos\theta_{\text{lion}}, R\sin\theta_{\text{lion}}) \\ \infty & \text{otherwise} \end{cases}$$

greatly penalizes the zebra (minimizer) for being caught. □

LECTURE 15

One-Player Dynamic Games

This lecture considers one-player discrete-time dynamic games, i.e., the optimal control of a discrete-time dynamical system. In subsequent lectures, we shall see that the computation of saddle-point equilibria can be reduced to the solution of one-player dynamic games, because saddle-point equilibria are defined through two coupled (single-player) optimizations.

15.1 One-Player Discrete-Time Games

We start by discussing solution methods for one-player dynamic games, which are simple optimizations. In the context of discrete-time dynamic games, this corresponds to dynamics of the form

$$\underbrace{x_{k+1}}_{\substack{\text{state at}\\ \text{stage } k+1}} = \underbrace{f_k}_{\substack{\text{"dynamics" at}\\ \text{stage } k}} \Big(\underbrace{x_k}_{\substack{\text{state at}\\ \text{stage } k}}, \underbrace{u_k}_{\substack{\text{P}_1\text{'s action}\\ \text{at stage } k}} \Big), \quad \forall k \in \{1, 2, \ldots, K\}, \tag{15.1}$$

starting at some initial state x_1 in the state space $\mathcal{X}$. At each time k, the action u_k is required to belong to a given action space $\mathcal{U}_k$. We assume finite horizon ($K < \infty$) stage-additive costs of the form

$$J := \sum_{k=1}^{K} g_k(x_k, u_k) \tag{15.2}$$

that the (only) player wants to minimize using either an open-loop policy

$$u_k = \gamma_k^{\text{OL}}(x_1), \quad \forall k \in \{1, 2, \ldots, K\},$$

or a state-feedback policy

$$u_k = \gamma_k^{\text{FB}}(x_k), \quad \forall k \in \{1, 2, \ldots, K\}.$$

15.2 Discrete-Time Cost-To-Go

Suppose that the player is at some state x at stage ℓ. This state would perhaps not be the optimal place to be at this stage, but nevertheless the player would like to estimate the cost if playing optimally from this point on so as to minimize the costs incurred in the remaining stages.

This scenario motivates defining the *cost-to-go from state $x \in \mathcal{X}$ at time $\ell \in$* $\{1, 2, \ldots, K\}$ by

Notation. The cost-to-go is a function of both x and ℓ and is often called the *value function* of the game/optimization.

$$V_\ell(x) := \inf_{u_\ell \in \mathcal{U}_\ell,\, u_{\ell+1} \in \mathcal{U}_{\ell+1},\, \ldots,\, u_K \in \mathcal{U}_K} \sum_{k=\ell}^{K} g_k(x_k, u_k), \quad \forall x \in \mathcal{X}, \tag{15.3}$$

with the sequence $\{x_k \in \mathcal{X} : k = \ell, \ell + 1, \ldots, K\}$ starting at

$$x_\ell = x$$

and satisfying the dynamics

$$x_{k+1} = f_k(x_k, u_k), \quad \forall k \in \{\ell, \ell + 1, \ldots, K - 1\}.$$

Computing the cost-to-go $V_1(x_1)$ from the initial state x_1 at the first stage $\ell = 1$ essentially amounts to minimizing the cost (15.2) for the dynamics (15.1). This observation leads to two important conclusions:

1. Regardless of the information structure considered (open loop, state feedback, or other), it is not possible to obtain a cost (15.2) lower than $V_1(x_1)$. This is because in the minimization in (15.3) we place no constraints on what information may or may not be available to compute the optimal u_k.

Note. Therefore $V_1(x_1)$ is always a lower bound on the smallest value that can be achieved for (15.2).

2. If the infimum in (15.3) is achieved for some specific sequence

$$u_1^* \in \mathcal{U}_1, \quad u_2^* \in \mathcal{U}_1, \quad \ldots, \quad u_K^* \in \mathcal{U}_K$$

that can be computed before the game starts just with knowledge of x_1, then this sequence of actions provides an optimal *open-loop* policy γ^{OL}:

Note. This would not be the case, e.g., if there were stochastic events to consider.

$$\gamma_1^{\mathrm{OL}}(x_1) := u_1^*, \quad \gamma_2^{\mathrm{OL}}(x_1) := u_2^*, \quad \ldots, \quad \gamma_K^{\mathrm{OL}}(x_1) := u_K^*.$$

In this case, $V_1(x_1)$ is precisely equal to the smallest value that can be achieved for (15.2).

15.3 Discrete-Time Dynamic Programming

Dynamic programming is a computationally efficient recursive technique that can be used to compute the cost-to-go. For the last stage K, the cost-to-go $V_K(x)$ is simply the minimum of

$$g_K(x_K, u_K)$$

Note. When $g_K(\cdot)$ is continuously differentiable, the optimization in (15.4) can be done using calculus by solving $\frac{dg_K(x_K,u_K)}{du_K}=0$. Recall that the global minima to a continuously differentiable function on a set $\mathcal{U}_K$ can only occur at points where the derivative is zero or at the boundary of $\mathcal{U}_K$.

over the possible actions u_K, for a game that starts with $x_K = x$ and therefore

$$V_K(x) = \inf_{u_K \in \mathcal{U}_K} g_K(x, u_K), \quad \forall x \in \mathcal{X}. \tag{15.4}$$

Therefore, for each state x, we can compute $V_K(x)$ by solving a single-parameter optimization over the set $\mathcal{U}_K$. For the previous stages $\ell < K$, we have that

$$\begin{aligned} V_\ell(x) &:= \inf_{u_\ell \in \mathcal{U}_\ell, \dots, u_K \in \mathcal{U}_K} \sum_{k=\ell}^{K} g_k(x_k, u_k) \\ &= \inf_{u_\ell \in \mathcal{U}_\ell, \dots, u_K \in \mathcal{U}_K} \Big(\underbrace{g_\ell(x, u_\ell)}_{\substack{\text{independent of} \\ u_{\ell+1}, \dots, u_K}} + \underbrace{\sum_{k=\ell+1}^{K} g_k(x_k, u_k)}_{\substack{\text{dependent on all} \\ u_\ell, \dots, u_K}} \Big) \\ &= \inf_{u_\ell \in \mathcal{U}_\ell} \Big(g_\ell(x, u_\ell) + \inf_{u_{\ell+1} \in \mathcal{U}_{\ell+1}, \dots, u_K \in \mathcal{U}_K} \sum_{k=\ell+1}^{K} g_k(x_k, u_k) \Big), \end{aligned} \tag{15.5}$$

where in the first equality we used the fact that we must set $x_\ell = x$ to compute $V_\ell(x)$ and in the second we used the fact that $g_\ell(x, u_\ell)$ does not depend on $u_{\ell+1}, \dots, u_K$. However,

$$\inf_{u_{\ell+1} \in \mathcal{U}_{\ell+1}, \dots, u_K \in \mathcal{U}_K} \sum_{k=\ell+1}^{K} g_k(x_k, u_k)$$

is precisely the minimum cost for a game starting at stage $\ell + 1$ with the state

$$x_{\ell+1} = f_\ell(x, u_\ell),$$

which is precisely the cost-to-go $V_{\ell+1}(f_\ell(x, u_\ell))$. We therefore conclude that

$$V_\ell(x) = \inf_{u_\ell \in \mathcal{U}_\ell} \big(g_\ell(x, u_\ell) + V_{\ell+1}(f_\ell(x, u_\ell)) \big), \quad \forall x \in \mathcal{X}, \quad \ell \in \{1, 2, \dots, K-1\}. \tag{15.6}$$

This shows that if we know the function $V_{\ell+1}(\cdot)$, then we can compute each $V_\ell(x)$ by solving a single-parameter optimization over the set $\mathcal{U}_\ell$. Moreover, this optimization, produces the optimal action u_ℓ^* to be used when the state is at x_ℓ.

It is convenient to define

$$V_{K+1}(x) = 0, \quad \forall x \in \mathcal{X},$$

which allows us to re-write both (15.4) and (15.6) using the formula:

$$V_\ell(x) = \inf_{u_\ell \in \mathcal{U}_\ell} \big(g_\ell(x, u_\ell) + V_{\ell+1}(f_\ell(x, u_\ell)) \big), \quad \forall x \in \mathcal{X}, \tag{15.7}$$

that is now valid $\forall \ell \in \{1, 2, \dots, K\}$.

For the particular case of $\ell = 1$ and $x = x_1$ and when the infima in (15.5) are actually minima, the points at which these infima are achieved can be used to construct an open-loop policy. Specifically, we can obtain:

- u_1^* from (15.7) with $\ell = 1$ and $x = x_1$, leading to $x_2^* = f_1(x_1, u_1^*)$;
- u_2^* from (15.7) with $\ell = 2$ and $x = x_2^*$, leading to $x_3^* = f_2(x_2^*, u_2^*)$;
- u_3^* from (15.7) with $\ell = 3$ and $x = x_3^*$, etc.

15.3.1 Open-Loop Optimization

The observations above justify the following procedure to compute the *open-loop* policy γ^{OL} that minimizes the cost (15.2) for the dynamics (15.1):

Attention! To do the backwards iteration, we need to compute each $V_\ell(x)$ for every possible value of the state x at stage ℓ.

MATLAB® Hint 6. For games with finite state spaces and finite action spaces, the backwards iteration in (15.8) can be implemented very efficiently in MATLAB®. ▹ p. 186

1. Compute the cost-to-go using a *backward iteration* starting from $\ell = K$ and proceeding backward in time until $\ell = 1$ using

$$V_{K+1}(x) = 0, \quad V_\ell(x) = \inf_{u_\ell \in \mathcal{U}_\ell} \left(g_\ell(x, u_\ell) + V_{\ell+1}(f_\ell(x, u_\ell)) \right), \quad \forall x \in \mathcal{X}. \tag{15.8}$$

2. Compute the sequence of actions

$$u_1^* \in \mathcal{U}_1, u_2^* \in \mathcal{U}_1, \ldots, u_K^* \in \mathcal{U}_K,$$

that minimize $V_1(x_1)$ using a *forward iteration* starting from $k = 1$ and proceeding forward in time until $k = K$:

$$x_1^* = x_1, \quad u_k^* = \arg\min_{u_k \in \mathcal{U}_k} \underbrace{\Big(g_k(x_k^*, u_k) + V_{k+1}(f_k(x_k^*, u_k)) \Big)}_{\text{computed using the precomputed states } x_k^*}, \quad x_{k+1}^* = f_k(x_k^*, u_k^*). \tag{15.9}$$

In (15.9), we are assuming that the infimum of $g_\ell(x_\ell^*, u_\ell) + V_{\ell+1}(f_\ell(x_\ell^*, u_\ell))$ is achieved at some point $u_k \in \mathcal{U}_k$. If this is not the case, then this procedure fails. When the infimum is achieved at multiple points, any one can be used in (15.9).

3. Finally, the optimal open-loop policy γ^{OL} is given by

$$\gamma_1^{OL}(x_1) := u_1^*, \quad \gamma_2^{OL}(x_1) := u_2^*, \quad \ldots, \quad \gamma_K^{OL}(x_1) := u_K^*. \tag{15.10}$$

Note. All the x_k^* and u_k^* in (15.9) are precomputed and depend solely on the initial state x_1, thus (15.10) is indeed an open-loop policy.

15.3.2 State-Feedback Optimization

Suppose that one uses the optimal open-loop policy γ^{OL} defined by (15.10), which selects the actions

$$u_k = \gamma_k^{OL}(x_1) := u_k^*, \quad \forall k \in \{1, 2, \ldots, K\}.$$

In this case, the precomputed states x_k^* defined by (15.9) match precisely the states x_k that would be measured during the game. Therefore we would get precisely the same

Note. In (15.11), we are assuming that the infimum of $g_\ell(x_\ell, u_\ell) + V_{\ell+1}(f_\ell(x_\ell, u_\ell))$ is achieved at some point $u_k \in \mathcal{U}_k$. If this is not the case, then this procedure fails. When the infimum is achieved at multiple points, any one can be used in (15.11).

minimum value $V_1(x_1)$ for the cost (15.2), if we were using a state-feedback policy γ^{FB} defined by

$$\gamma_k^{\text{FB}}(x_k) := \underbrace{\arg\min_{u_k \in \mathcal{U}_k} \Big(g_k(x_k, u_k) + V_{k+1}(f_k(x_k, u_k))\Big)}_{\text{computed using the measured state } x_k}, \quad \forall k \in \{1, 2, \dots, K\}. \tag{15.11}$$

When all the $g_k(x_k, u_k) + V_{k+1}\big(f_k(x_k, u_k)\big)$ have a minimum for some $u_k \in \mathcal{U}_k$, this state-feedback policy γ^{FB} can do as well as the optimal open-loop policy γ^{OL}. Since it is not possible to obtain a value for (15.2) lower than $V_1(x_1)$, we conclude that (15.11) is an optimal state-feedback policy.

Notation 5 (Time-consistent policy). A state-feedback policy such as (15.11) that minimizes the cost-to-go from the current state x_k at time k is said to be *time consistent*.

There may be state-feedback policies $\bar{\gamma}^{\text{FB}}$ that still achieve a cost as low as $V_1(x_1)$, but are not time consistent because $\bar{\gamma}_k^{\text{FB}}(x_k)$ may not achieve the minimum in (15.11) for every state x_k. This is possible because it is irrelevant for a policy to achieve the minimum in (15.11) for states x_k that are never reached through an optimal path. Nevertheless, time-consistent policies are more robust in the sense that, if due to an unexpected event the state at some time k is taken to a point other than

Note. Open-loop policies are also not robust because they rely on precomputed states and cannot react to unexpected events such as (15.12).

$$x_{k+1} \neq f_k(x_k, u_k), \tag{15.12}$$

then a time-consistent policy is still optimal in minimizing the cost-to-go from the stage $k+1$ forward. □

The following result summarizes our conclusions above for open-loop and state-feedback:

Theorem 15.1. Consider the sequence of functions $V_1(x), V_2(x), \dots, V_{K+1}(x)$ uniquely defined by

$$V_k(x) = \begin{cases} 0 & k = K+1 \\ \inf_{u_k \in \mathcal{U}_k} \Big(g_k(x, u_k) + V_{k+1}(f_k(x, u_\ell))\Big) & k \in \{1, 2, \dots, K\} \end{cases} \quad \forall x \in \mathcal{X}. \tag{15.13}$$

Then $V_k(x)$ is equal to the cost-to-go and if the infimum in (15.13) is always achieved at some point in $\mathcal{U}_k$, we have that:

1. For any given initial state x_1, an optimal open-loop policy γ^{OL} is given by

$$\gamma_k^{\text{OL}}(x_1) := u_k^*, \quad \forall k \in \{1, 2, \dots, K\},$$

Note. In an open-loop setting both x_k^* and u_k^*, $\forall k \in \{1, 2, \dots, K\}$ are precomputed before the game starts.

with u_k^* obtained from solving

$$x_1^* = x_1, \quad u_k^* = \underbrace{\arg\min_{u_k \in \mathcal{U}_k} \Big(g_k(x_k^*, u_k) + V_{k+1}(f_k(x_k^*, u_k))\Big)}_{\text{computed using the precomputed states } x_k^*}, \quad x_{k+1}^* = f_k(x_k^*, u_k^*).$$

Notation 5. A state-feedback policy such as (15.14) that minimizes the cost-to-go is said to be *time consistent*.
▹ p. 182

2. An optimal (time-consistent) state-feedback policy γ^{FB} is given by

$$\gamma_k^{\text{FB}}(x_k) := \arg\min_{u_k \in \mathcal{U}_k} \underbrace{\Big(g_k(x_k, u_k) + V_{k+1}(f_k(x_k, u_k))\Big)}_{\text{computed using the measured state } x_k}, \quad \forall k \in \{1, 2, \ldots, K\}. \tag{15.14}$$

Note 13. Open-loop and state-feedback information structures are "optimal," in the sense that it is not possible to achieve a cost lower than $V_1(x_1)$, regardless of the information structure.
▹ p. 184

Note. Both the open-loop and the state-feedback policies lead precisely to the same trajectory, so we are able to treat them jointly in this proof.

Either of the above optimal policies leads to an optimal cost equal to $V_1(x_1)$. □

Although this result was essentially already proved, the following (compact) proof provides additional insight and will help us to prove the analogous result in continuous time.

Proof of Theorem 15.1. Let u_k^* and x_k^*, $\forall k \in \{1, 2, \ldots, K\}$ be a trajectory arising from either the open-loop or the state-feedback policies and let $\bar{u}_k, \bar{x}_k$, $\forall k \in \{1, 2, \ldots, K\}$ be another (arbitrary) trajectory. To prove optimality, we need to show that the latter trajectory cannot lead to a cost lower than the former.

Since $V_k(x)$ satisfies (15.13) and u_k^* achieves the infimum in (15.13), for every $k \in \{1, 2, \ldots, K\}$

$$V_k(x_k^*) = \inf_{u_k \in \mathcal{U}_k} \Big(g_k(x_k^*, u_k) + V_{k+1}(f_k(x_k^*, u_k))\Big) = g_k(x_k^*, u_k^*) + V_{k+1}(f_k(x_k^*, u_k^*)). \tag{15.15}$$

However, since $\bar{u}_k$ does not necessarily achieve the infimum, we have that

$$V_k(\bar{x}_k) = \inf_{u_k \in \mathcal{U}_k} \Big(g_k(\bar{x}_k, u_k) + V_{k+1}(f_k(\bar{x}_k, u_k))\Big) \le g_k(\bar{x}_k, \bar{u}_k) + V_{k+1}(f_k(\bar{x}_k, \bar{u}_k)). \tag{15.16}$$

Summing both sides of (15.15) from $k = 1$ to $k = K$, we conclude that

$$\sum_{k=1}^{K} V_k(x_k^*) = \sum_{k=1}^{K} g_k(x_k^*, u_k^*) + \sum_{k=1}^{K} V_{k+1}\Big(\underbrace{f_k(x_k^*, u_k^*)}_{x_{k+1}^*}\Big)$$

$$\Leftrightarrow \sum_{k=1}^{K} V_k(x_k^*) - \sum_{k=1}^{K} V_{k+1}(x_{k+1}^*) = \sum_{k=1}^{K} g_k(x_k^*, u_k^*).$$

Since

$$\sum_{k=1}^{K} V_k(x_k^*) - \sum_{k=1}^{K} V_{k+1}(x_{k+1}^*) = V_1(x_1) - V_{K+1}(x_{K+1}^*) = V_1(x_1),$$

we conclude that

$$V_1(x_1) = \sum_{k=1}^{K} g_k(x_k^*, u_k^*). \tag{15.17}$$

Now summing both sides of (15.16) from $k = 1$ to $k = K$, we similarly conclude that

$$\sum_{k=1}^{K} V_k(\bar{x}_k) \leq \sum_{k=1}^{K} g_k(\bar{x}_k, \bar{u}_k) + \sum_{k=1}^{K} V_{k+1}\Big(\underbrace{f_k(\bar{x}_k, \bar{u}_k)}_{\bar{x}_{k+1}}\Big)$$

$$\Leftrightarrow \sum_{k=1}^{K} V_k(\bar{x}_k) - \sum_{k=1}^{K} V_{k+1}(\bar{x}_{k+1}) \leq \sum_{k=1}^{K} g_k(\bar{x}_k, \bar{u}_k) \;\Leftrightarrow\; V_1(x_1) \leq \sum_{k=1}^{K} g_k(\bar{x}_k, \bar{u}_k),$$

from which we obtain

$$V_1(x_1) = \sum_{k=1}^{K} g_k(x_k^*, u_k^*) \leq \sum_{k=1}^{K} g_k(\bar{x}_k, \bar{u}_k). \tag{15.18}$$

Two conclusions can be drawn from (15.17) and (15.18):

1. The signal $\bar{u}_k$ does not lead to a cost that is smaller than that of u_k^*.
2. $V_1(x_1)$ is equal to the optimal cost obtained with u_k^*.

If we had carried out the above proof on an interval $\{\ell, \ell+1, \ldots, K\}$ with initial state $x_\ell = x$, we would have concluded that $V_\ell(x)$ is the (optimal) value of the cost-to-go from state x at time ℓ. ■

Note 13. In the proof of Theorem 15.1 we actually showed that it is not possible to achieve a cost lower than $V_1(x_1)$, regardless of the information structure. This is because the signal $\bar{u}_k$ considered could have been generated by a policy using any information structure and we showed that $\bar{u}_k$ cannot lead to a cost smaller than $V_1(x_1)$. □

15.4 Computational Complexity

For large state-spaces $\mathcal{X}$, the computational effort needed to compute the cost-to-go at all stages can be very large. One may then wonder if it is worth using dynamic programming, instead of doing an exhaustive search. To decide which option is best one needs to estimate the computation involved in exploring each option. We shall do this assuming finite state-spaces and finite action spaces.

Exhaustive Search. Suppose that a game has K stages and that at the stage ℓ the number of actions available to the player is equal to $|\mathcal{U}_\ell|$. An exhaustive search over all possible selections of actions requires comparing the costs associated with as many options as

$$|\mathcal{U}_1| \times |\mathcal{U}_2| \times \cdots \times |\mathcal{U}_K|. \tag{15.19}$$

Dynamic Programming. At a particular stage ℓ and for a specific value of the state x, computing the cost-to-go $V_\ell(x)$ requires comparing all the actions available, which

TABLE 15.1 Computation complexity of solving the one-player Tic-Tac-Toe game in Example 15.1 using dynamic programming.

Stage	Number of $\times$'s	Number of $\circ$'s	$\lvert\mathcal{X}_\ell\rvert$	$\lvert\mathcal{U}_\ell\rvert$	$\lvert\mathcal{X}_\ell\rvert \times \lvert\mathcal{U}_\ell\rvert$
1	0	0	1	9	9
2	1	0	9	8	72
3	1	1	$9 \times 8 = 72$	7	504
4	2	1	$\binom{9}{2} \times 7 = 252$	6	1512
5	2	2	$\binom{9}{2} \times \binom{7}{2} = 756$	5	3780
6	3	2	$\binom{9}{3} \times \binom{6}{2} = 1260$	4	5040
7	3	3	$\binom{9}{3} \times \binom{6}{3} = 1680$	3	5040
8	4	3	$\binom{9}{4} \times \binom{5}{3} = 1260$	2	2520
9	4	4	$\binom{9}{4} \times \binom{5}{4} = 630$	1	630
10	5	4	$\binom{9}{5} = 126$	0	0
			Total number of comparisons needed		19107

roughly requires making $|\mathcal{U}_\ell|$ comparisons. Since this has to be done for every state x and for every stage $\ell \in \{1, 2, \ldots, K\}$, the total number of comparisons is roughly equal to

$$|\mathcal{U}_1| \times |\mathcal{X}_1| + |\mathcal{U}_2| \times |\mathcal{X}_2| + \cdots + |\mathcal{U}_K| \times |\mathcal{X}_K|, \tag{15.20}$$

where we are denoting by $|\mathcal{X}_\ell|$ the total number of possible states at the stage ℓ. By comparing (15.19) with (15.20) we see that dynamic programming can result in significant savings provided that the size of the state space is small when compared to (15.19).

Example 15.1 (Tic-Tac-Toe). Consider a (silly) version of the Tic-Tac-Toe game in which the same player places all the marks. An *exhaustive search* among all possible ways to play would have to consider 9 possible ways to place the first $\times$, 8 possible ways to place the subsequent $\circ$, 7 possible ways to place the first $\times$, etc., leading to a total of

$$9! = 9 \times 8 \times \cdots \times 2 \times 1 = 362880$$

distinct options that must be compared. For *dynamic programming*, the total number of comparisons needed is computed in Table 15.1 and turns out to be about 19 times smaller than what would be needed for an exhaustive search. In larger games, the difference between the two approaches is even more spectacular. Essentially, this happens because many different sequences of actions collapse to the same state. □

15.5 Solving Finite One-Player Games with MATLAB®

For games with finite state spaces and finite action spaces, the backwards iteration in (15.8) can be implemented very efficiently in MATLAB®. To this effect, suppose that we enumerate all states so that the state-space can be viewed as

$$\mathcal{X} := \{1, 2, \ldots, n_{\mathcal{X}}\}$$

and that we enumerate all actions so that the action space can be viewed as

$$\mathcal{U} := \{1, 2, \ldots, n_{\mathcal{U}}\}.$$

For simplicity, we shall assume that all states can occur at every stage and that all actions are also available at every stage.

In this case, each function $f_k(x, u)$ and $g_k(x, u)$ that define the game dynamics and the stage-cost, respectively, can be represented by an $n_{\mathcal{X}} \times n_{\mathcal{U}}$ matrix with one row per state and one column per action. On the other hand, each cost-to-go $V_k(x)$ can be represented by an $n_{\mathcal{X}} \times 1$ column vector with one row per state. Suppose then that the following variables have been constructed within MATLAB®:

Note. Recall that we enumerated the states and actions so that the elements of $\mathcal{X}$ and $\mathcal{U}$ are positive integers.

- `F` is a cell-array with K elements, each equal to an $n_{\mathcal{X}} \times n_{\mathcal{U}}$ matrix so that `F{k}` represents the game dynamics function $f_k(x, u)$, $\forall x \in \mathcal{X}, u \in \mathcal{U}, k \in \{1, 2, \ldots, K\}$. Specifically, the entry `F{k}(i,j)` of the matrix `F{k}` is the state $f_k(i, j)$.
- `G` is a cell-array with K elements, each equal to an $n_{\mathcal{X}} \times n_{\mathcal{U}}$ matrix so that `G{k}` represents the stage-cost function $g_k(x, u)$, $\forall x \in \mathcal{X}, u \in \mathcal{U}, k \in \{1, 2, \ldots, K\}$. Specifically, the entry `G{k}(i,j)` of the matrix `G{k}` is the per-stage cost $g_k(i, j)$.

With these definitions, we can construct the cost-to-go $V_k(x)$ in (15.13) very efficiently using the following MATLAB® code:

MATLAB® Hint 1. The arguments `[],2` in the `min` function indicate that the minimization should be performed along the second dimension of the matrix (i.e., along the columns). ▷ p. 27

```
V{K+1}=zeros(size(G{K},1),1);
for k=K:-1:1
  V{k}=min(G{k}+V{k+1}(F{k}),[],2);
end
```

After running this code, the following variable has been created:

- `V` is a cell-array with $K + 1$ elements, each equal to an $n_{\mathcal{X}} \times 1$ column vector so that `V{k}` represents the cost-to-go $V_k(x)$, $\forall x \in \mathcal{X}, k \in \{1, 2, \ldots, K\}$. Specifically, the entry `V{k}(i)` of the vector `V{k}` is the cost-to-go $V_k(i)$ from state `i` at state `k`.

For a given state x at stage k, the optimal action u given by (15.14) can be obtained using

```
[~,u]=min(G{k}(x,:)+V{k+1}(F{k}(x,:))',[],2);
```

15.6 Linear Quadratic Dynamic Games

Discrete-time *linear quadratic one-player games* are characterized by linear dynamics of the form

$$x_{k+1} = \underbrace{Ax_k + Bu_k}_{f_k(x_k,u_k)}, \quad x \in \mathbb{R}^n, \quad u \in \mathbb{R}^{n_u}, \quad k \in \{1, 2, \ldots, K\}$$

and a stage-additive quadratic cost of the form

$$J := \sum_{k=1}^{K} \underbrace{\left(\|y_k\|^2 + u_k' R u_k\right)}_{g_k(x_k,u_k)} = \sum_{k=1}^{K} \underbrace{\left(x_k' C' C x_k + u_k' R u_k\right)}_{g_k(x_k,u_k)},$$

where

$$y_k = Cx_k, \quad \forall k \in \{1, 2, \ldots, K\}.$$

This cost function captures scenarios in which the (only) player wants to make the y_k, $k \in \{1, 2, \ldots, K\}$ small without "spending" much effort in their action u_k. The symmetric positive definite matrix R can be seen as a conversion factor that maps units of u_k into units of y_k. Theorem 15.1 can be used to compute optimal policies for this game and leads to the following result.

Note. See Exercise 15.1. ▷ p. 187

Notation. The equation (15.21) is called a *difference Riccati equation.*

Corollary 15.1. Suppose that we define the matrices P_k according to the following (backwards) recursion:

$$P_{K+1} = 0, \tag{15.21a}$$

$$P_k = C'C + A'P_{k+1}A - A'P_{k+1}B(R + B'P_{k+1}B)^{-1}B'P_{k+1}A, \quad \forall k \in \{1, 2, \ldots, K\}, \tag{15.21b}$$

and that

$$R + B'P_{k+1}B \geq 0, \quad \forall k \in \{1, 2, \ldots, K\}. \tag{15.22}$$

Then the state-feedback policy

$$\gamma_k^{FB}(x_k) = -(R + B'P_{k+1}B)^{-1}B'P_{k+1}Ax_k, \quad \forall k \in \{1, 2, \ldots, K\}$$

is an optimal (time-consistent) state-feedback policy for the linear quadratic one-player game, leading to an optimal cost equal to $x_1'P_1x_1$. □

15.7 Practice Exercise

15.1. Prove Corollary 15.1.

Hint: Try to find a solution to (15.13) of the form $V_k(x) = x'P_kx$, $\forall x \in \mathbb{R}^n$, $k \in \{1, 2, \ldots, K+1\}$ *for appropriately selected symmetric* $n \times n$ *matrices* P_k.

Solution to Exercise 15.1. For this game, (15.13) is given by

$$V_k(x) = \begin{cases} 0 & k = K+1 \\ \min_{u_k \in \mathbb{R}^{n_u}} (x'C'Cx + u_k' R u_k + V_{k+1}(Ax + Bu_k)) & k \in \{1, 2, \dots, K\}, \end{cases} \tag{15.23}$$

$\forall x \in \mathbb{R}^n$. Inspired by the quadratic form of the per-stage cost, we will try to find a solution to (15.23) of the form

$$V_k(x) = x' P_k x, \quad \forall x \in \mathbb{R}^n, \ k \in \{1, 2, \dots, K+1\},$$

for appropriately selected symmetric $n \times n$ matrices P_k. For (15.23) to hold, we need to have $P_{K+1} = 0$ and

$$\begin{aligned} x'P_k x &= \min_{u_k \in \mathbb{R}^{n_u}} \left(x'C'Cx + u_k' R u_k + (Ax + Bu_k)' P_{k+1} (Ax + Bu_k) \right) \\ &= \min_{u_k \in \mathbb{R}^{n_u}} \left(x'(C'C + A'P_{k+1}A)x + u_k'(R + B'P_{k+1}B)u_k + 2x'A'P_{k+1}Bu_k \right), \end{aligned} \tag{15.24}$$

$\forall x \in \mathbb{R}^n$, $k \in \{1, 2, \dots, K\}$. Since the function to optimize is quadratic, to compute the minimum in (15.24) we simply need to make the appropriate gradient equal to zero:

Note. The critical point obtained by setting $\frac{\partial(\cdot)}{\partial u_k} = 0$ is indeed a minimum because of (15.22).

Note. The value for the minimum will provide the optimal policy.

$$\begin{aligned} &\frac{\partial}{\partial u_k} \left(x'(C'C + A'P_{k+1}A)x + u_k'(R + B'P_{k+1}B)u_k + 2x'A'P_{k+1}Bu_k \right) = 0 \\ &\Leftrightarrow \ 2u_k'(R + B'P_{k+1}B) + 2x'A'P_{k+1}B = 0 \\ &\Leftrightarrow \ u_k = -(R + B'P_{k+1}B)^{-1} B'P_{k+1}Ax. \end{aligned} \tag{15.25}$$

Therefore

$$\begin{aligned} &\min_{u_k \in \mathbb{R}^{n_u}} \Big(\underbrace{x'(C'C + A'P_{k+1}A)x + u_k'(R + B'P_{k+1}B)u_k + 2x'A'P_{k+1}Bu_k}_{u_k = -(R + B'P_{k+1}B)^{-1}B'P_{k+1}Ax} \Big) \\ &\qquad = x'(C'C + A'P_{k+1}A - A'P_{k+1}B(R + B'P_{k+1}B)^{-1}B'P_{k+1}A)x. \end{aligned}$$

This means that (15.24) is of the form

$$x'P_k x = x'(C'C + A'P_{k+1}A - A'P_{k+1}B(R + B'P_{k+1}B)^{-1}B'P_{k+1}A)x,$$

which holds in view of (15.21). Corollary 15.1 then follows directly from Theorem 15.1, since we have found a sequence of functions $V_1(x), V_2(x), \dots, V_{K+1}(x)$ that satisfies (15.13) for which the infimum is always achieved at the point u_k given by (15.25).

15.8 Additional Exercise

15.2. Prove the following result, which permits the construction of a state-feedback policy based on a sequence of functions that satisfies (15.13) only approximately.

Note. For $\epsilon = 0$, the equation (15.26) is equivalent to equation (15.13) and, if we also have $\delta = 0$, (15.27) shows that the infimum is achieved at $u_k = \gamma(x_k)$. In this case, (15.28) with $\epsilon = \delta = 0$ means that no other policy $\bar{\gamma}$ can lead to a cost lower than that of γ, which we already knew to be so from Theorem 15.1. With $\delta > 0$ and $\epsilon > 0$, there may exist policies $\bar{\gamma}$ that improve upon γ, but never by more than $(2\epsilon + \delta)K$, which can be quite small if ϵ and δ are both very small.

Theorem 15.2. Suppose that there exist constants $\epsilon, \delta \geq 0$; a sequence of functions $V_1(x), V_2(x), \ldots, V_{K+1}(x)$ for which

$$V_{K+1}(x) = 0, \quad \left| V_k(x) - \inf_{u_k \in \mathcal{U}_k} \big(g_k(x, u_k) + V_{k+1}(f_k(x, u_k))\big) \right| \leq \epsilon, \quad \forall k \in \{1, 2, \ldots, K\}, \tag{15.26}$$

$\forall x \in \mathcal{X}$; and a state-feedback policy $\gamma(\cdot)$ for which

$$g_k(x, u_k) + V_{k+1}(f_k(x, u_k))\Big|_{u_k = \gamma(x)} \leq \delta + \inf_{u_k \in \mathcal{U}_k} g_k(x, u_k) + V_{k+1}(f_k(x, u_k)), \tag{15.27}$$

$\forall x \in \mathcal{X}, \;\; k \in \{1, 2, \ldots, K\}$. Then the policy $\gamma(\cdot)$ leads to a cost $J(\gamma)$ that satisfies

$$J(\gamma) \leq J(\bar{\gamma}) + (2\epsilon + \delta)K, \tag{15.28}$$

for any other state-feedback policy $\bar{\gamma}(\cdot)$ with cost $J(\bar{\gamma})$. □

LECTURE 16

One-Player Differential Games

This lecture considers one-player continuous-time dynamic games, i.e., the optimal control of a continuous-time dynamical system.

16.1 One-Player Continuous-Time Differential Games

Consider now a one-player continuous-time differential game with dynamics of the form

$$\underbrace{\dot{x}(t)}_{\substack{\text{state}\\\text{derivative}}} = \underbrace{f}_{\substack{\text{game}\\\text{dynamics}}}\Big(\underbrace{t}_{\text{time}}, \underbrace{x(t)}_{\substack{\text{current}\\\text{state}}}, \underbrace{u(t)}_{\substack{\text{P}_1\text{'s action}\\\text{at time } t}}\Big), \quad \forall t \in [0, T], \tag{16.1}$$

with state $x(t) \in \mathbb{R}^n$ initialized at a given $x(0) = x_0$. For every time $t \in [0, T]$, the action $u(t)$ is required to belong to a given action space $\mathcal{U}$. We assume a finite horizon ($T < \infty$) integral cost of the form

$$J := \underbrace{\int_0^T g(t, x(t), u(t))dt}_{\text{cost along trajectory}} + \underbrace{q(x(T))}_{\text{final cost}} \tag{16.2}$$

that the (only) player wants to minimize either using an open-loop policy

$$u(t) = \gamma^{\text{OL}}(t, x_0), \quad \forall t \in [0, T],$$

or a state-feedback policy

$$u(t) = \gamma^{\text{FB}}(t, x(t)), \quad \forall t \in [0, T].$$

16.2 Continuous-Time Cost-To-Go

The definition of cost-to-go for differential games follows the same scenario used for discrete-time games: A player is at some state x at time τ and wants to estimate the cost if playing optimally so as to minimize the cost incurred from this point forwards until the end of the game.

Notation. The cost-to-go is a function of both x and τ and is often called the *value function* of the game/optimization.

Formally, the *cost-to-go from state x at time τ* by

$$V(\tau, x_\tau) := \inf_{u(t)\in\mathcal{U}, \forall t\in[\tau,T]} \int_\tau^T g(t, x(t), u(t))dt + q(x(T)) \tag{16.3}$$

with the state $x(t)$, $t \in [\tau, T]$ initialized at

$$x(\tau) = x_\tau$$

and satisfying the dynamics

$$\dot{x}(t) = f(t, x(t), u(t)), \quad \forall t \in [\tau, T].$$

Computing the cost-to-go $V(0, x_0)$ from the initial state x_0 at time $\tau = 0$ essentially amounts to minimizing the cost (16.2) for the dynamics (16.1). This observation leads to two important conclusions:

1. Regardless of the information structure considered (open loop, state feedback, or other), it is not possible to obtain a cost (16.2) lower than $V(0, x_0)$. This is because in the minimization in (16.3) we place no constraints on what information may or may not be available to compute the optimal $u(t)$, $\forall t \in [\tau, T]$.
2. If the infimum in (15.3) is achieved for some specific signal

$$u^*(t) \in \mathcal{U}, \quad t \in [\tau, T]$$

Note. This would not be case, e.g., if there were stochastic events to consider.

that can be computed before the game starts just with knowledge of x_0, then this action signal provides an optimal *open-loop* policy γ^{OL}:

$$\gamma_1^{\mathrm{OL}}(t, x_0) := u^*(t), \quad \forall t \in [\tau, T].$$

16.3 Continuous-Time Dynamic Programming

Also in continuous-time it is possible to compute the cost-to-go somewhat recursively. For the final time T, the cost-to-go $V(T, x_T)$ is simply given by

$$V(T, x_T) = q(x(T)) = q(x_T),$$

because for $\tau = T$ the integral term in (16.3) disappears and the game starts (and ends) precisely at $x(T) = x_T$.

Consider now some time $\tau < T$ and pick some small positive constant h so that $\tau + h$ is still smaller than T. Then

$$\begin{aligned} V(\tau, x_\tau) &= \inf_{u(t)\in\mathcal{U}, \forall t\in[\tau,T]} \int_\tau^T g(t, x(t), u(t))dt + q(x(T)) \\ &= \inf_{u(t)\in\mathcal{U}, \forall t\in[\tau,T]} \underbrace{\int_\tau^{\tau+h} g(t, x(t), u(t))dt}_{\text{independent of } u(t), t\in[\tau+h,T]} + \underbrace{\int_{\tau+h}^T g(t, x(t), u(t))dt + q(x(T))}_{\text{depends on all } u(t), t\in[\tau,T]} \\ &= \inf_{u(t)\in\mathcal{U}, \forall t\in[\tau,\tau+h)} \Bigg(\int_\tau^{\tau+h} g(t, x(t), u(t))dt \\ &\quad + \inf_{u(t)\in\mathcal{U}, \forall t\in[\tau+h,T]} \int_{\tau+h}^T g(t, x(t), u(t))dt + q(x(T)) \Bigg). \end{aligned}$$

Recognizing that the "inner" infimum is precisely the cost-to-go from the state $x(\tau + h)$ at time $\tau + h$, we can re-write these equations compactly as

$$V(\tau, x_\tau) = \inf_{u(t)\in\mathcal{U},\ \forall t\in[\tau,\tau+h)} \left(\int_\tau^{\tau+h} g(t, x(t), u(t))dt + V(\tau + h, x(\tau + h)) \right).$$

Subtracting $V(\tau, x_\tau) = V(\tau, x(\tau))$ from both sides and dividing both sides by $h > 0$, we can further re-write the above equation as

$$0 = \inf_{u(t)\in\mathcal{U},\ \forall t\in[\tau,\tau+h)} \left(\frac{1}{h} \int_\tau^{\tau+h} g(t, x(t), u(t))dt + \frac{V(\tau + h, x(\tau + h)) - V(\tau, x(\tau))}{h} \right). \tag{16.4}$$

Since the left hand side must be equal to zero for every $h \in (0, T - \tau)$, the limit of the right hand side as $h \to 0$ must also be equal to zero. If we optimistically assume that the limit of the infimum is the same as the infimum of the limit and also that all limits exist, we could use the following equalities

Attention! Since it is unclear whether or not this is a reasonable assumption, we will later have to confirm that any conclusions reached are indeed valid.

$$\lim_{h\to 0} \frac{1}{h} \int_\tau^{\tau+h} g(t, x(t), u(t))dt = g(\tau, x(\tau), u(\tau))$$

$$\begin{aligned} &\lim_{h\to 0} \frac{V(\tau + h, x(\tau + h)) - V(\tau, x(\tau))}{h} \\ &\quad = \frac{dV(\tau, x(\tau))}{d\tau} = \frac{\partial V(\tau, x(\tau))}{\partial \tau} + \frac{\partial V(\tau, x(\tau))}{\partial x} f(\tau, x(\tau), u(\tau)) \end{aligned}$$

to transform (16.4) into the so-called *Hamilton-Jacobi-Bellman (HJB)* equation:

$$0 = \inf_{u\in\mathcal{U}} \left(g(\tau, x, u) + \frac{\partial V(\tau, x)}{\partial \tau} + \frac{\partial V(\tau, x)}{\partial x} f(\tau, x, u) \right), \quad \forall \tau \in [0, T], x \in \mathbb{R}^n. \tag{16.5}$$

Note. The infimum in the equation (16.4) is taken over all the values of the signal $u(t)$ in the interval $t \in [\tau, \tau + h)$, which is the subject of calculus of variations. However, in the HJB equation (16.5) the infimum is simply taken over the set $\mathcal{U}$ and can generally be solved using standard calculus.

It turns out that this equation is indeed quite useful to compute the cost-to-go.

Theorem 16.1 (Hamilton-Jacobi-Bellman). Any continuously differentiable function $V(\tau, x)$ that satisfies the Hamilton-Jacobi-Bellman equation (16.5) with

$$V(T, x) = q(x), \quad \forall x \in \mathbb{R}^n \tag{16.6}$$

is equal to the cost-to-go $V(\tau, x)$. In addition, if the infimum in (16.5) is always achieved at some point in $\mathcal{U}$, we have that:

1. For any given x_0, an optimal open-loop policy γ^{OL} is given by

$$\gamma^{\mathrm{OL}}(t, x_0) := u^*(t), \quad \forall t \in [0, T]$$

with $u^*(t)$ obtained from solving

$$u^*(t) = \arg\min_{u\in\mathcal{U}} g(t, x^*(t), u) + \frac{\partial V(t, x^*(t))}{\partial x} f(t, x^*(t), u)$$

$$\dot{x}^*(t) = f(t, x^*(t), u^*(t)), \quad \forall t \in [0, T], \;\; x^*(0) = x_0.$$

2. An optimal (time-consistent) state-feedback policy γ^{FB} is given by

$$\gamma^{\mathrm{FB}}(t, x(t)) := \arg\min_{u\in\mathcal{U}} g(t, x(t), u) + \frac{\partial V(t, x(t))}{\partial x} f(t, x(t), u), \quad \forall t \in [0, T]. \tag{16.7}$$

Either of the above optimal policies leads to an optimal cost equal to $V(0, x_0)$. □

Note 14. The HJB equation (16.5) is a partial differential equation (PDE) and (16.6) can be viewed as a boundary condition for this PDE. ▹ p. 193

Note. In an open-loop setting both $x^*(t)$ and $u^*(t), t \in [0, T]$ are precomputed before the game starts.

Notation 5. A state-feedback policy such as (16.7) that minimizes the cost-to-go is said to be *time consistent*. ▹ p. 182

Note 15. Open-loop and state-feedback information structures are "optimal," in the sense that it is not possible to achieve a cost lower than $V(0, x_0)$, regardless of the information structure. ▹ p. 194

Note 14 (Hamilton-Jacobi-Bellman equation). Since $\frac{\partial V(\tau,x)}{\partial x}$ in (16.5) does not depend of u, we can remove this term from inside the infimum, which leads to the following common form for the HJB equation:

$$-\frac{\partial V(\tau, x)}{\partial \tau} = \inf_{u\in\mathcal{U}}\Big(g(\tau, x, u) + \frac{\partial V(\tau, x)}{\partial x} f(\tau, x, u)\Big).$$

This form highlights the fact that the Hamilton-Jacobi-Bellman is a partial differential equation (PDE) and we can view (16.6) as a boundary condition for this PDE.

When we can find a continuously differentiable solution to this PDE that satisfies the appropriate boundary condition, we automatically obtain the cost-to-go. Unfortunately, solving a PDE is often difficult to solve and many times (16.5) does not have continuously differentiable solutions. □

Note. The lack of differentiability of the solution to (16.5) is not an insurmountable difficulty, as there are methods to overcome this technical difficulty by making sense of what it means for a non-differentiable function to be a solution to (16.5) [2].

Note. Both the open-loop and the state feedback policies lead precisely to the same trajectory.

Proof of Theorem 16.1. Let $u^*(t)$ and $x^*(t)$, $\forall t \in [0, T]$ be a trajectory arising from either the open-loop or the state-feedback policies and let $\bar{u}(t)$, $\bar{x}(t)$, $\forall t \in [0, T]$ be another (arbitrary) trajectory. To prove optimality, we need to show that the latter trajectory cannot lead to a cost lower than the former.

Since $V(\tau, x)$ satisfies the Hamilton-Jacobi-Bellman equation (16.5) and $u^*(t)$ achieves the infimum in (16.5), for every $t \in [0, T]$, we have that

$$\begin{aligned} 0 &= \inf_{u \in \mathcal{U}} g(t, x^*(t), u) + \frac{\partial V(t, x^*(t))}{\partial t} + \frac{\partial V(t, x^*(t))}{\partial x} f(t, x^*(t), u) \\ &= g(t, x^*(t), u^*(t)) + \frac{\partial V(t, x^*(t))}{\partial t} + \frac{\partial V(t, x^*(t))}{\partial x} f(t, x^*(t), u^*(t)). \end{aligned} \tag{16.8}$$

However, since $\bar{u}(t)$ does not necessarily achieve the infimum, we have that

$$\begin{aligned} 0 &= \inf_{u \in \mathcal{U}} g(t, \bar{x}(t), u) + \frac{\partial V(t, \bar{x}(t))}{\partial t} + \frac{\partial V(t, \bar{x}(t))}{\partial x} f(t, \bar{x}(t), u) \\ &\leq g(t, \bar{x}(t), \bar{u}(t)) + \frac{\partial V(t, \bar{x}(t))}{\partial t} + \frac{\partial V(t, \bar{x}(t))}{\partial x} f(t, \bar{x}(t), \bar{u}(t)). \end{aligned} \tag{16.9}$$

Integrating both sides of (16.8) and (16.9) over the interval $[0, T]$, we conclude that

$$\begin{aligned} 0 &= \int_0^T \Big(g(t, x^*(t), u^*(t)) + \underbrace{\frac{\partial V(t, x^*(t))}{\partial t} + \frac{\partial V(t, x^*(t))}{\partial x} f(t, x^*(t), u^*(t))}_{\frac{dV(t, x^*(t))}{dt}} \Big) dt \\ &\leq \int_0^T \Big(g(t, \bar{x}(t), \bar{u}(t)) + \underbrace{\frac{\partial V(t, \bar{x}(t))}{\partial t} + \frac{\partial V(t, \bar{x}(t))}{\partial x} f(t, \bar{x}(t), \bar{u}(t))}_{\frac{dV(t, \bar{x}(t))}{dt}} \Big) dt, \end{aligned}$$

from which we obtain

$$\begin{aligned} 0 &= \int_0^T g(t, x^*(t), u^*(t)) dt + V(T, x^*(T)) - V(0, x_0) \\ &\leq \int_0^T g(t, \bar{x}(t), \bar{u}(t) dt + V(T, \bar{x}(T)) - V(0, x_0). \end{aligned}$$

Using (16.6) and adding $V(0, x_0)$ to all terms, one concludes that

$$V(0, x_0) = \int_0^T g(t, x^*(t), u^*(t)) dt + q(x^*(T)) \leq \int_0^T g(t, \bar{x}(t), \bar{u}(t)) dt + q(\bar{x}(T)).$$

Two conclusions can be drawn from here: First, the signal $\bar{u}(t)$ does not lead to a cost smaller than that of $u^*(t)$ and, second, $V(0, x_0)$ is equal to the optimal cost obtained with $u^*(t)$. If we had carried out the above proof on an interval $[\tau, T]$ with initial state $x(\tau) = x$, we would have concluded that $V(\tau, x)$ is the (optimal) value of the cost-to-go from state x at time τ. ■

Note 15. In the proofs of Theorems 16.1 (and later in the proof of Theorem 16.2) we actually show that it is not possible to achieve a cost lower than $V(0, x_0)$, regardless of the information structure. This is because the signal $\bar{u}(t)$ considered could have been generated by a policy using any information structure and we showed $\bar{u}(t)$ cannot lead to a cost smaller than $V(0, x_0)$. □

16.4 Linear Quadratic Dynamic Games

Continuous-time *linear quadratic one-player games* are characterized by linear dynamics of the form

$$\dot{x} = \underbrace{Ax(t) + Bu(t)}_{f(t,x(t),u(t))}, \quad x \in \mathbb{R}^n, \quad u \in \mathbb{R}^{n_u}, \quad t \in [0, T]$$

and an integral quadratic cost of the form

$$J := \int_0^T \underbrace{\Big(\|y(t)\|^2 + u(t)'Ru(t)\Big)}_{g(t,x(t),u(t))} dt + \underbrace{x'(T)P_T x(T)}_{q(x(T))},$$

where

$$y(t) = Cx(t), \quad \forall t \in [0, T].$$

This cost function captures scenarios in which the (only) player wants to make $y(t)$ small over the interval $[0, T]$ without "spending" much effort in the action $u(t)$. The symmetric positive definite matrix R can be seen as a conversion factor that maps units of $u(t)$ into units of $y(t)$.

The Hamilton-Jacobi-Bellman equation (16.5) for this game is given by

$$-\frac{\partial V(t, x)}{\partial t} = \min_{u \in \mathbb{R}^{n_u}} \Big(x'C'Cx + u'Ru + \frac{\partial V(t, x)}{\partial x}(Ax + Bu)\Big), \tag{16.10}$$

$\forall x \in \mathbb{R}^n, \quad t \in [0, T]$, with

$$V(T, x) = x'P_T x, \quad \forall x \in \mathbb{R}^n. \tag{16.11}$$

Inspired by the boundary condition (16.11), we will try to find a solution to (16.10) of the form

$$V(t, x) = x'P(t)x, \quad \forall x \in \mathbb{R}^n, \quad t \in [0, T]$$

for some appropriately selected symmetric $n \times n$ matrix $P(t)$. For (16.11) to hold, we need to have $P(T) = P_T$ and, for (16.10) to hold, we need

$$-x'\dot{P}(t)x = \min_{u \in \mathbb{R}^{n_u}} \Big(x'C'Cx + u'Ru + 2x'P(t)(Ax + Bu)\Big),$$

$$\forall x \in \mathbb{R}^n, \quad t \in [0, T]. \tag{16.12}$$

Note. The critical point obtained by setting $\frac{\partial(\cdot)}{\partial u} = 0$ is indeed a minimum because $R > 0$ in the quadratic form in (16.12).

Note. The value for the minimum will provide the optimal policy.

Since the function to optimize is quadratic, to compute the minimum in (16.12) we simply need to make the appropriate gradient equal to zero:

$$\frac{\partial}{\partial u}\Big(x'C'Cx + u'Ru + 2x'P(Ax + Bu)\Big) = 0$$

$$\Leftrightarrow \quad 2u'R + 2x'PB = 0 \quad \Leftrightarrow \quad u = -R^{-1}B'Px.$$

Note. Since P is symmetric, we can write $2x'PAx$ as $x'(PA + A'P)x$.

Therefore

$$\min_{u \in \mathbb{R}^{n_u}} \Big(\underbrace{x'C'Cx + u'Ru + 2x'P(Ax + Bu)}_{u=-R^{-1}B'Px} \Big)$$
$$= x'(PA + A'P + C'C - PBR^{-1}B'P)x.$$

This means that (16.12) is of the form

$$-x'\dot{P}(t)x = x'(PA + A'P + C'C - PBR^{-1}B'P)x, \quad \forall x \in \mathbb{R}^n, \ t \in [0, T],$$

which holds provided that

$$-\dot{P}(t) = PA + A'P + C'C - PBR^{-1}B'P, \quad \forall t \in [0, T].$$

Theorem 16.1 can then be used to compute the optimal policies for this game and leads to the following result.

Note. The function $P(t)$ could be found by numerically solving the matrix-valued ordinary differential equation (16.13) backwards in time.

Notation. The equation (16.13) is called a *differential Riccati equation.*

Corollary 16.1. Suppose that there exists a symmetric solution to the following matrix-valued ordinary differential equation

$$-\dot{P}(t) = PA + A'P + C'C - PBR^{-1}B'P, \quad \forall t \in [0, T], \tag{16.13}$$

with final condition $P(T) = P_T$. Then the state-feedback policy

$$\gamma^*(t, x) = -R^{-1}B'Px, \quad \forall x \in \mathbb{R}^n, \ t \in [0, T]$$

is an optimal state-feedback policy, leading to an optimal cost equal to $x_0'P(0)x_0$. □

16.5 Differential Games with Variable Termination Time

Consider now a one-player continuous-time differential game with the usual dynamics

$$\dot{x}(t) = f(t, x(t), u(t)), \quad x(t) \in \mathbb{R}^n, \ u(t) \in \mathcal{U}, \ t \geq 0,$$

and initialized at a given $x(0) = x_0$, but with an integral cost with variable horizon

$$J := \underbrace{\int_0^{T_{\text{end}}} g(t, x(t), u(t))dt}_{\text{cost along trajectory}} + \underbrace{q(T_{\text{end}}, x(T_{\text{end}}))}_{\text{final cost}},$$

Notation. The states in $\mathcal{X}_{\text{end}}$ are often called the *game-over* states.

where T_{end} is the first time at which the state $x(t)$ enters a closed set $\mathcal{X}_{\text{end}} \subset \mathbb{R}^n$ or $T_{\text{end}} = +\infty$ in case $x(t)$ never enters $\mathcal{X}_{\text{end}}$. We can think of $\mathcal{X}_{\text{end}}$ as the set of states at which the game terminates, as the evolution of $x(t)$ is irrelevant after this time.

Notation. The cost-to-go is a function of both x and τ and is often called the *value function* of the game/optimization.

For this game, the *cost-to-go from state x at time τ* is defined by

$$V(\tau, x) := \inf_{u(t) \in \mathcal{U}, \forall t \geq \tau} \int_\tau^{T_{\text{end}}} g(t, x(t), u(t))dt + q(T_{\text{end}}, x(T_{\text{end}})),$$

where the state $x(t)$, $t \geq \tau$ satisfies the dynamics

$$x(\tau) = x, \quad \dot{x}(t) = f(t, x(t), u(t)), \quad \forall t \geq \tau$$

and T_{end} denotes the first time at which $x(t)$ enters the closed set $\mathcal{X}_{\text{end}}$.

When we compute $V(\tau, x)$ for some $x \in \mathcal{X}_{\text{end}}$, we have $T_{\text{end}} = \tau$ and therefore

$$V(\tau, x) = q(\tau, x), \quad \forall \tau \geq 0, \quad x \in \mathcal{X}_{\text{end}},$$

instead of the boundary condition (16.6). However, it turns out that the Hamilton-Jacobi-Bellman equation (16.5) is still the same and we have the following result:

Note 14. The HJB equation (16.5) is a partial differential equation (PDE) and (16.14) can be viewed as a boundary condition for this PDE. ▹ p. 193

Theorem 16.2. Any continuously differentiable function $V(\tau, x)$ that satisfies the Hamilton-Jacobi-Bellman equation (16.5) with

$$V(\tau, x) = q(\tau, x), \quad \forall \tau \geq 0, \quad x \in \mathcal{X}_{\text{end}}, \tag{16.14}$$

is equal to the cost-to-go $V(\tau, x)$. In addition, if the infimum in (16.5) is always achieved at some point in $\mathcal{U}$, we have that:

1. For any given x_0, an optimal open-loop policy γ^{OL} is given by

$$\gamma^{\text{OL}}(t, x_0) := u^*(t), \quad \forall t \in [0, T_{\text{end}}]$$

Note. In an open-loop setting both $x^*(t)$ and $u^*(t)$, $t \in [0, T_{\text{end}}]$ are precomputed before the game starts.

with $u^*(t)$ obtained from solving

$$u^*(t) = \arg\min_{u \in \mathcal{U}} g(t, x^*(t), u) + \frac{\partial V(t, x^*(t))}{\partial x} f(t, x^*(t), u)$$

$$\dot{x}^*(t) = f(t, x^*(t), u^*(t)), \quad \forall t \in [0, T_{\text{end}}], \quad x^*(0) = x_0.$$

Notation 5. A state-feedback policy such as (16.15) that minimizes the cost-to-go is said to be *time consistent*. ▹ p. 182

2. An optimal (time-consistent) state-feedback policy γ^{FB} is given by

$$\gamma^{\text{OL}}(t, x(t)) := \arg\min_{u \in \mathcal{U}} g(t, x(t), u) + \frac{\partial V(t, x(t))}{\partial x} f(t, x(t), u), \quad \forall t \in [0, T_{\text{end}}]. \tag{16.15}$$

Note 15. Open-loop and state-feedback information structures are "optimal," in the sense that it is not possible to achieve a cost lower than $V(0, x_0)$, regardless of the information structure. ▹ p. 194

Either of the above optimal policies leads to an optimal cost equal to $V(0, x_0)$. □

Proof of Theorem 16.2. Let $u^*(t)$ and $x^*(t)$, $\forall t \geq 0$ be a trajectory arising from either the open-loop or the state-feedback policies and let $\bar{u}(t)$, $\bar{x}(t)$, $\forall t \geq 0$ be another (arbitrary) trajectory. To prove optimality, we need to show that the latter trajectory cannot lead to a cost lower than the former.

Note. Both the open-loop and the state-feedback policies lead precisely to the same trajectory.

Proceeding exactly as in the proof of Theorem 16.1, we conclude that since $V(\tau, x)$ satisfies the Hamilton-Jacobi-Bellman equation (16.5) and $u^*(t)$ achieves the infimum in (16.5), equations (16.8) and (16.9) hold. Let $T^*_{\text{end}} \in [0, \infty]$ and $\bar{T}_{\text{end}} \in [0, \infty]$ denote the times at which $x^*(t)$ and $\bar{x}(t)$, respectively, enter the set $\mathcal{X}_{\text{end}}$. Integrating both sides of (16.8) and (16.9) over the intervals $[0, T^*_{\text{end}}]$ and $[0, \bar{T}_{\text{end}}]$, respectively, we conclude that

$$0 = \int_0^{T^*_{\text{end}}} \Big(g(t, x^*(t), u^*(t)) + \underbrace{\frac{\partial V(t, x^*(t))}{\partial t} + \frac{\partial V(t, x^*(t))}{\partial x} f(t, x^*(t), u^*(t))}_{\frac{dV(t, x^*(t))}{dt}} \Big) dt$$

$$\leq \int_0^{\bar{T}_{\text{end}}} \Big(g(t, \bar{x}(t), \bar{u}(t)) + \underbrace{\frac{\partial V(t, \bar{x}(t))}{\partial t} + \frac{\partial V(t, \bar{x}(t))}{\partial x} f(t, \bar{x}(t), \bar{u}(t))}_{\frac{dV(t, \bar{x}(t))}{dt}} \Big) dt,$$

from which we obtain

$$0 = \int_0^{T^*_{\text{end}}} g(t, x^*(t), u^*(t))dt + V(T^*_{\text{end}}, x^*(T^*_{\text{end}})) - V(0, x_0)$$

$$\leq \int_0^{\bar{T}_{\text{end}}} g(t, \bar{x}(t), \bar{u}(t)dt + V(\bar{T}_{\text{end}}, \bar{x}(\bar{T}_{\text{end}})) - V(0, x_0).$$

Using (16.14), two conclusions can be drawn from here: First, the signal $\bar{u}(t)$ does not lead to a cost smaller than that of $u^*(t)$, because

$$\int_0^{T^*_{\text{end}}} g(t, x^*(t), u^*(t))dt + q(T^*_{\text{end}}, x^*(T^*_{\text{end}}))$$

$$\leq \int_0^{\bar{T}_{\text{end}}} g(t, \bar{x}(t), \bar{u}(t)dt + q(\bar{T}_{\text{end}}, \bar{x}(\bar{T}_{\text{end}})).$$

Second, $V(0, x_0)$ is equal to the optimal cost obtained with $u^*(t)$, because

$$V(0, x_0) = \int_0^{T^*_{\text{end}}} g(t, x^*(t), u^*(t))dt + q(T^*_{\text{end}}, x^*(T^*_{\text{end}})).$$

If we had carried out the above proof starting at time τ with initial state $x(\tau) = x$, we would have concluded that $V(\tau, x)$ is the (optimal) value of the cost-to-go from state x at time τ. ■

16.6 Practice Exercise

16.1. Prove the following result, which permits the construction of a state-feedback policy based on a function that only satisfies the Hamilton-Jacobi-Bellman equation (16.5) approximately.

Note. For $\epsilon = 0$, the equation (16.16) is equivalent to the Hamilton-Jacobi-Bellman equation (16.5) and, if we also have $\delta = 0$, (16.18) shows that the infimum is achieved at $u_k = \gamma(x_k)$. In this case, (16.19) with $\epsilon = \delta = 0$ means that no other policy $\bar{\gamma}$ can lead to a cost lower than that of γ, which we already knew to be truth from Theorem 16.1. With $\delta > 0$ and $\epsilon > 0$, there may exist policies $\bar{\gamma}$ that improve upon γ, but never by more than $(2\epsilon + \delta)T$, which can be quite small if ϵ and δ are both very small.

Theorem 16.3. Suppose that there exist constants ϵ, $\delta \geq 0$, a continuously differentiable function $V(\tau, x)$ that satisfies

$$\left| \inf_{u \in \mathcal{U}} \left(g(\tau, x, u) + \frac{\partial V(\tau, x)}{\partial \tau} + \frac{\partial V(\tau, x)}{\partial x} f(\tau, x, u) \right) \right| \leq \epsilon, \quad \forall \tau \in [0, T], \quad x \in \mathbb{R}^n \tag{16.16}$$

with

$$V(T, x) = q(x), \quad \forall x \in \mathbb{R}^n, \tag{16.17}$$

and a state-feedback policy $\gamma(\cdot)$ for which

$$\begin{aligned} g(\tau, x, u) + \frac{\partial V(\tau, x)}{\partial x} f(\tau, x, u) \Big|_{u=\gamma(x)} \\ \leq \delta + \inf_{u \in \mathcal{U}} \left(g(\tau, x, u) + \frac{\partial V(\tau, x)}{\partial x} f(\tau, x, u) \right), \end{aligned} \tag{16.18}$$

$\forall \tau \in [0, T], \quad x \in \mathbb{R}^n$. Then the policy $\gamma(\cdot)$ leads to a cost $J(\gamma)$ that satisfies

$$J(\gamma) \leq J(\bar{\gamma}) + (2\epsilon + \delta)T, \tag{16.19}$$

for any other state-feedback policy $\bar{\gamma}(\cdot)$ with cost $J(\bar{\gamma})$. □

Solution to Exercise 16.1 and Proof of Theorem 16.3. Let $u^*(t)$ and $x^*(t)$, $\forall t \in [0, T]$ be a trajectory arising from the state-feedback policy $\gamma(\cdot)$ and let $\bar{u}(t), \bar{x}(t), \forall t \in [0, T]$ be another trajectory, e.g., resulting from the state-feedback policy $\bar{\gamma}(\cdot)$ that appears in (16.19).

Since $V(\tau, x)$ satisfies (16.16) for $x = x^*(t)$ and $\gamma(\cdot)$ satisfies (16.18) for $x = x^*(t)$ and $u = \gamma(x^*(t)) = u^*(t)$ satisfies (16.18), we have that

$$\begin{aligned} \epsilon &\geq \inf_{u \in \mathcal{U}} \left(g(t, x^*(t), u) + \frac{\partial V(t, x^*(t))}{\partial t} + \frac{\partial V(t, x^*(t))}{\partial x} f(t, x^*(t), u) \right) \\ &\geq -\delta + g(t, x^*(t), u^*(t)) + \frac{\partial V(t, x^*(t))}{\partial t} + \frac{\partial V(t, x^*(t))}{\partial x} f(t, x^*(t), u^*(t)). \end{aligned} \tag{16.20}$$

On the other hand, using (16.16) for $x = \bar{x}(t)$ and the fact that $\bar{u}(t)$ does not necessarily achieve the infimum, we have that

$$\begin{aligned} -\epsilon &\leq \inf_{u \in \mathcal{U}} \left(g(t, \bar{x}(t), u) + \frac{\partial V(t, \bar{x}(t))}{\partial t} + \frac{\partial V(t, \bar{x}(t))}{\partial x} f(t, \bar{x}(t), u) \right) \\ &\leq g(t, \bar{x}(t), \bar{u}(t)) + \frac{\partial V(t, \bar{x}(t))}{\partial t} + \frac{\partial V(t, \bar{x}(t))}{\partial x} f(t, \bar{x}(t), \bar{u}(t)). \end{aligned} \tag{16.21}$$

Integrating both sides of (16.20) and (16.21) over the interval $[0, T]$, we conclude that

$$
\begin{aligned}
(\epsilon + \delta)T &\geq \int_0^T \Big(g(t, x^*(t), u^*(t)) + \underbrace{\frac{\partial V(t, x^*(t))}{\partial t} + \frac{\partial V(t, x^*(t))}{\partial x} f(t, x^*(t), u^*(t))}_{\frac{dV(t, x^*(t))}{dt}}\Big)dt \\
&= \int_0^T g\Big(t, x^*(t), u^*(t))dt + V(T, x^*(T)\Big) - V(0, x_0)
\end{aligned}
$$

and

$$
\begin{aligned}
-\epsilon T &\leq \int_0^T \Big(g(t, \bar{x}(t), \bar{u}(t)) + \underbrace{\frac{\partial V(t, \bar{x}(t))}{\partial t} + \frac{\partial V(t, \bar{x}(t))}{\partial x} f(t, \bar{x}(t), \bar{u}(t))}_{\frac{dV(t, \bar{x}(t))}{dt}}\Big)dt \\
&= \int_0^T g\Big(t, \bar{x}(t), \bar{u}(t)\Big)dt + V(T, \bar{x}(T)) - V(0, x_0),
\end{aligned}
$$

from which we obtain

$$
\begin{aligned}
&\int_0^T g(t, x^*(t), u^*(t))dt + V(T, x^*(T)) \\
&\qquad \leq V(0, x_0) + (\epsilon + \delta)T \leq (2\epsilon + \delta)T + \int_0^T g(t, \bar{x}(t), \bar{u}(t))dt + V(T, \bar{x}(T)),
\end{aligned}
$$

which proves (16.19).

LECTURE 17

State-Feedback Zero-Sum Dynamic Games

This lecture addresses the computation of saddle-point equilibria of zero-sum discrete-time dynamic games in state-feedback policies.

17.1 Zero-Sum Dynamic Games in Discrete Time

We now discuss solution methods for two-player zero-sum dynamic games in discrete time, which correspond to dynamics of the form

$$\underbrace{x_{k+1}}_{\substack{\text{entry node at}\\ \text{stage } k+1}} = \underbrace{f_k}_{\substack{\text{"dynamics" at}\\ \text{stage } k}} \Big(\underbrace{x_k}_{\substack{\text{entry node}\\ \text{at stage } k}}, \underbrace{u_k}_{\substack{\text{P}_1\text{'s action}\\ \text{at stage } k}}, \underbrace{d_k}_{\substack{\text{P}_2\text{'s action}\\ \text{at stage } k}} \Big), \quad \forall k \in \{1, 2, \dots, K\},$$

starting at some initial state x_1 in the state space $\mathcal{X}$. At each time k, P_1's action u_k is required to belong to an action space $\mathcal{U}_k$ and P_2's action d_k is required to belong to an action space $\mathcal{D}_k$. We assume a finite horizon ($K < \infty$) stage-additive cost of the form

$$\sum_{k=1}^{K} g_k(x_k, u_k, d_k), \tag{17.1}$$

that P_1 wants to minimize and P_2 wants to maximize. In this section we consider a *state-feedback information structure*, which corresponds to policies of the form

$$u_k = \gamma_k(x_k), \quad d_k = \sigma_k(x_k), \quad \forall k \in \{1, 2, \dots, K\}.$$

Suppose that for a given state-feedback policy γ for P_1 and a given state-feedback policy σ for P_2, we denote by $J(\gamma, \sigma)$ the corresponding value of the cost (17.1). Our goal is to find a saddle-point pair of equilibrium policies (γ^*, σ^*) for which

$$J(\gamma^*, \sigma) \le J(\gamma^*, \sigma^*) \le J(\gamma, \sigma^*), \quad \forall \gamma \in \Gamma_1, \ \sigma \in \Gamma_2, \tag{17.2}$$

where Γ_1 and Γ_2 denote the sets of all state-feedback policies for P_1 and P_2, respectively. Re-writing (17.2) as

$$J(\gamma^*, \sigma^*) = \min_{\gamma \in \Gamma_1} J(\gamma, \sigma^*), \qquad J(\gamma^*, \sigma^*) = \max_{\sigma \in \Gamma_2} J(\gamma^*, \sigma),$$

we conclude that if σ^* was known we could obtain γ^* from the following single-player optimization:

$$\begin{aligned} &\text{minimize over } \gamma \in \Gamma_1 \text{ the cost} && J(\gamma, \sigma^*) := \sum_{k=1}^{K} g_k(x_k, u_k, \sigma_k^*(x_k)) \\ &\text{subject to the dynamics} && x_{k+1} = f_k(x_k, u_k, \sigma_k^*(x_k)). \end{aligned}$$

In view of what we saw in Lecture 15, an optimal state-feedback policy γ^* could be constructed first using a backward iteration to compute the cost-to-go $V_k^1(x)$ for P_1 using

$$V_{K+1}^1(x) = 0,\ V_k^1(x) = \inf_{u_k \in \mathcal{U}_k} \Big(g_k(x, u_k, \sigma_k^*(x)) + V_{k+1}^1(f_k(x, u_k, \sigma_k^*(x))) \Big),$$
$$\forall k \in \{1, 2, \dots, K\}, \tag{17.3}$$

and then

$$\gamma_k^*(x) := \arg\min_{u_k \in \mathcal{U}_k} \Big(g_k(x, u_k, \sigma_k^*(x)) + V_{k+1}(f_k(x, u_k, \sigma_k^*(x))) \Big)$$
$$\forall k \in \{1, 2, \dots, K\}. \tag{17.4}$$

Moreover, the minimum $J(\gamma^*, \sigma^*)$ is given by $V_1^1(x_1)$.

Note. We only derived the dynamic programming equations for single-player minimizations, but for single-player maximizations analogous formulas are still valid, provided that we replace all infima by suprema.

Similarly, if γ^* was known we could obtain an optimal state-feedback policy σ^* from the following single-player optimization:

$$\begin{aligned} &\text{maximize over } \sigma \in \Gamma_2 \text{ the reward} && J(\gamma^*, \sigma) := \sum_{k=1}^{K} g_k(x_k, \gamma_k^*(x_k), d_k) \\ &\text{subject to the dynamics} && x_{k+1} = f_k(x_k, \gamma_k^*(x_k), d_k). \end{aligned}$$

In view of what we saw in Lecture 15, such σ^* could be constructed first using a backward iteration to compute the cost-to-go $V_k^2(x)$ for P_2 using

$$V_{K+1}^2(x) = 0,\ V_k^2(x) = \sup_{d_k \in \mathcal{D}_k} \Big(g_k(x, \gamma_k^*(x), d_k) + V_{k+1}^2(f_k(x, \gamma_k^*(x), d_k)) \Big),$$
$$\forall k \in \{1, 2, \dots, K\}, \tag{17.5}$$

and then

$$\sigma_k^*(x) := \arg\max_{d_k \in \mathcal{D}_k} \Big(g_k(x, \gamma_k^*(x), d_k) + V_{k+1}(f_k(x, \gamma_k^*(x), d_k)) \Big),$$
$$\forall k \in \{1, 2, \dots, K\}. \tag{17.6}$$

Moreover, the maximum $J(\gamma^*, \sigma^*)$ is given by $V_1^2(x_1)$.

17.2 Discrete-Time Dynamic Programming

The key to finding the saddle-point pair of equilibrium policies (γ^*, σ^*) is to realize that it is possible to construct a pair of state-feedback policies for which the four equations (17.3), (17.4), (17.5), (17.6) all hold.

To see how this can be done, consider the costs-to-go V_K^1, V_K^2 and state-feedback policies γ_K^*, σ_K^* at the last stage. For (17.3), (17.4), (17.5), and (17.6) to hold we need that

$$V_K^1(x) = \inf_{u_K \in \mathcal{U}_K} g_K(x, u_K, \sigma_K^*(x)), \qquad \gamma_K^*(x) := \arg\min_{u_K \in \mathcal{U}_K} g_K(x, u_K, \sigma_K^*(x)),$$

$$V_K^2(x) = \sup_{d_K \in \mathcal{D}_K} g_K(x, \gamma_K^*(x), d_K), \qquad \sigma_K^*(x) := \arg\min_{d_K \in \mathcal{D}_K} g_K(x, \gamma_K^*(x), d_K),$$

which can be re-written equivalently as

$$V_K^1(x) = g_K(x, \gamma_K^*(x), \sigma_K^*(x)) \le g_K(x, u_K, \sigma_K^*(x)), \quad \forall u_K \in \mathcal{U}_K$$

$$V_K^2(x) = g_K(x, \gamma_K^*(x), \sigma_K^*(x)) \ge g_K(x, \gamma_K^*(x), d_K), \quad \forall d_K \in \mathcal{D}_K.$$

Since the right-hand side of the top and bottom equalities are the same, we conclude that $V_K^1(x) = V_K^2(x)$. Moreover, this shows that the pair $(\gamma_K^*(x), \sigma_K^*(x)) \in \mathcal{U}_K \times \mathcal{D}_K$ must be a saddle-point equilibrium for the zero-sum game with outcome

$$g_K(x, u_K, d_K)$$

and actions $u_K \in \mathcal{U}_K$ and $d_K \in \mathcal{D}_K$ for the minimizer and maximizer, respectively. Moreover, $V_K^1(x) = V_K^2(x)$ must be precisely equal to the value of this game. In view of the results that we saw for zero-sum games, this will only be possible if security policies exist and the security levels for both players are equal to the value of the game, i.e.,

Notation. By using a max and min (instead of sup and inf) in the outer optimizations, we implicitly require the outer sup and inf to be achieved at some point in the set, i.e., security policies do exist for this game.

$$V_K^1(x) = V_K^2(x) = V_K(x) := \min_{u_K \in \mathcal{U}_K} \sup_{d_K \in \mathcal{D}_K} g_K(x, u_K, d_K)$$

$$= \max_{d_K \in \mathcal{D}_K} \inf_{u_K \in \mathcal{U}_K} g_K(x, u_K, d_K).$$

Consider now the costs-to-go V_{K-1}^1, V_{K-1}^2 and state-feedback policies $\gamma_{K-1}^*, \sigma_{K-1}^*$ at stage $K-1$. For (17.3), (17.4), (17.5), and (17.6) to hold we need that

Note. We omit the superscripts in V_K^1 and V_K^2 in the right-hand side of these equations, since we have already seen that $V_K^1(x) = V_K^2(x)$.

$$V_{K-1}^1(x) = \inf_{u_{K-1} \in \mathcal{U}_{K-1}} \left(g_{K-1}(x, u_{K-1}, \sigma_{K-1}^*(x)) + V_K(f_{K-1}(x, u_{K-1}, \sigma_{K-1}^*(x))) \right),$$

$$\gamma_{K-1}^*(x) := \arg\min_{u_{K-1} \in \mathcal{U}_{K-1}} \left(g_{K-1}(x, u_{K-1}, \sigma_{K-1}^*(x)) + V_K(f_{K-1}(x, u_{K-1}, \sigma_{K-1}^*(x))) \right),$$

$$V_{K-1}^2(x) = \sup_{d_{K-1} \in \mathcal{D}_{K-1}} \left(g_{K-1}(x, \gamma_{K-1}^*(x), d_{K-1}) + V_K(f_{K-1}(x, \gamma_{K-1}^*(x), d_{K-1})) \right),$$

$$\sigma_{K-1}^*(x) := \arg\min_{d_{K-1} \in \mathcal{D}_{K-1}} \left(g_{K-1}(x, \gamma_{K-1}^*(x), d_{K-1}) + V_K(f_{K-1}(x, \gamma_{K-1}^*(x), d_{K-1})) \right).$$

So we now conclude that the pair $(\gamma^*_{K-1}(x), \sigma^*_{K-1}(x)) \in \mathcal{U}_{K-1} \times \mathcal{D}_{K-1}$ must be a saddle-point equilibrium for the zero-sum game with outcome

$$g_{K-1}(x, u_{K-1}, d_{K-1}) + V_K\Big(f_{K-1}(x, u_{K-1}, d_{K-1})\Big)$$

and actions $u_{K-1} \in \mathcal{U}_{K-1}$ and $d_{K-1} \in \mathcal{D}_{K-1}$ for the minimizer and maximizer, respectively. Moreover, $V^1_{K-1}(x) = V^2_{K-1}(x)$ must be precisely equal to the value of this game. Continuing this reasoning backwards in time all the way to the first stage, we obtain the following result:

Notation. By using a max and min (instead of sup and inf) in the outer optimizations in (17.7), we implicitly require the outer sup and inf to be achieved at some point in the sets.

Theorem 17.1. Assume that we can recursively compute functions $V_1(x)$, $V_2(x), \ldots, V_{K+1}(x)$, such that $\forall x \in \mathcal{X}, k \in \{1, 2, \ldots, K\}$, we have that

$$V_k(x) := \min_{u_k \in \mathcal{U}_k} \sup_{d_k \in \mathcal{D}_k} \Big(g_k(x, u_k, d_k) + V_{k+1}(f_k(x, u_k, d_k))\Big) \tag{17.7a}$$

$$= \max_{d_k \in \mathcal{D}_k} \inf_{u_k \in \mathcal{U}_k} \Big(g_k(x, u_k, d_k) + V_{k+1}(f_k(x, u_k, d_k))\Big), \tag{17.7b}$$

where

$$V_{K+1}(x) = 0, \quad \forall x \in \mathcal{X}. \tag{17.8}$$

MATLAB® Hint 7. For games with finite state spaces and finite action spaces, the backwards iteration in (17.7) can be implemented very efficiently in MATLAB®. ▷ p. 205

Then the pair of policies (γ^*, σ^*) defined as follows is a saddle-point equilibrium in state-feedback policies:

$$\gamma^*_k(x) := \arg\min_{u_k \in \mathcal{U}_k} \sup_{d_k \in \mathcal{D}_k} \Big(g_k(x, u_k, d_k) + V_{k+1}(f_k(x, u_k, d_k))\Big), \tag{17.9}$$

$$\sigma^*_k(x) := \arg\max_{d_k \in \mathcal{D}_k} \inf_{u_k \in \mathcal{U}_k} \Big(g_k(x, u_k, d_k) + V_{k+1}(f_k(x, u_k, d_k))\Big), \tag{17.10}$$

$\forall x \in \mathcal{X},\ \ k \in \{1, 2, \ldots, K\}$. Moreover, the value of the game is equal to $V_1(x_1)$. □

Attention! Theorem 17.1 provides a *sufficient condition* for the existence of Nash equilibria, but this condition is not necessary. In particular, the two security levels in (17.7) may not commute for some state x at some stage k, but there still may be a saddle-point for the game. We saw this before for games in extensive form.

Note. See Exercise 7.2. ▷ p. 86

When the min and max do not commute in (17.7) and $\mathcal{U}_k$ and $\mathcal{D}_k$ are finite, one may want to use a mixed saddle point, leading to behavioral policies (i.e., per-stage randomization).

Proof of Theorem 17.1. From the fact that the inf and sup commute in (17.7) and the definitions of $\gamma^*_k(x)$ and $\sigma^*_k(x)$, we conclude that the pair $(\gamma^*_k(x), \sigma^*_k(x))$ is a saddle-point equilibrium for a zero-sum game with criterion

$$\Big(g_k(x, u_k, d_k) + V_{k+1}(f_k(x, u_k, d_k))\Big),$$

which means that

$$g_k(x, \gamma_k^*(x), d_k) + V_{k+1}\Big(f_k(x, \gamma_k^*(x), d_k)\Big)$$
$$\leq g_k(x, \gamma_k^*(x), \sigma_k^*(x)) + V_{k+1}\Big(f_k(x, \gamma_k^*(x), \sigma_k^*(x))\Big)$$
$$\leq g_k(x, u_k, \sigma_k^*(x)) + V_{k+1}\Big(f_k(x, u_k, \sigma_k^*(x))\Big), \quad \forall u_k \in \mathcal{U}_k, d_k \in \mathcal{D}_k.$$

Moreover, since the middle term in these inequalities is also equal to the right-hand-side of (17.7), we have that

Note 16. We can actually conclude that P_2 cannot get a reward larger than $V_1(x_1)$ against $\gamma_k^*(x)$, regardless of the information structure available to P_2. ▷ p. 205

$$V_k(x) = g_k(x, \gamma_k^*(x), \sigma_k^*(x)) + V_{k+1}\Big(f_k(x, \gamma_k^*(x), \sigma_k^*(x))\Big)$$
$$= \sup_{d \in \mathcal{D}}\Big(g_k(x, \gamma_k^*(x), d) + V_{k+1}(f_k(x, \gamma_k^*(x), d))\Big), \quad \forall x \in \mathbb{R}^n, \quad t \in [0, T],$$

which, because of Theorem 15.1, shows that $\sigma_k^*(x)$ is an optimal (maximizing) state-feedback policy against $\gamma_k^*(x)$ and the maximum is equal to $V_1(x_1)$. Moreover, since we also have that

Note 16. We can actually conclude that P_1 cannot get a cost smaller than $V_1(x_1)$ against $\sigma_k^*(x)$, regardless of the information structure available to P_1. ▷ p. 205

$$V_k(x) = g_k(x, \gamma_k^*(x), \sigma_k^*(x)) + V_{k+1}\Big(f_k(x, \gamma_k^*(x), \sigma_k^*(x))\Big)$$
$$= \inf_{u \in \mathcal{U}}\Big(g_k(x, u, \sigma_k^*(x)) + V_{k+1}(f_k(x, u, \sigma_k^*(x)))\Big), \quad \forall x \in \mathbb{R}^n, \quad t \in [0, T],$$

we can also conclude that $\gamma_k^*(x)$ is an optimal (minimizing) state-feedback policy against $\sigma_k^*(x)$ and the minimum is also equal to $V_1(x_1)$. This proves that (γ^*, σ^*) is indeed a saddle-point equilibrium in state-feedback policies with value $V_1(x_1)$. ■

Note 16. In view of Note 13 (p. 184), we actually conclude that:

1. P_2 cannot get a reward larger than $V_1(x_1)$ against $\gamma_k^*(x)$, regardless of the information structure available to P_2, and
2. P_1 cannot get a cost smaller than $V_1(x_1)$ against $\sigma_k^*(x)$, regardless of the information structure available to P_1.

In practice, this means that $\gamma_k^*(x)$ and $\sigma_k^*(x)$ are "extremely safe" policies for P_1 and P_2, respectively, since they guarantee a level of reward regardless of the information structure for the other player. □

17.3 Solving Finite Zero-Sum Games with MATLAB®

For games with finite state spaces and finite action spaces, the backwards iteration in (17.7) can be implemented very efficiently in MATLAB®. To this effect, suppose that we enumerate all states so that the state-space can be viewed as

$$\mathcal{X} := \{1, 2, \ldots, n_{\mathcal{X}}\}$$

and that we enumerate all actions so that the action spaces can be viewed as

$$\mathcal{U} := \{1, 2, \ldots, n_{\mathcal{U}}\}, \qquad \mathcal{D} := \{1, 2, \ldots, n_{\mathcal{D}}\}.$$

For simplicity, we shall assume that all states can occur at every stage and that all actions are also available at every stage.

In this case, the functions $f_k(x, u, d)$ and $g_k(x, u, d)$ that define the game dynamics and the stage-cost, respectively, can be represented by a three-dimensional $n_{\mathcal{X}} \times n_{\mathcal{U}} \times n_{\mathcal{D}}$ tensor. On the other hand, each $V_k(x)$ can be represented by an $n_{\mathcal{X}} \times 1$ columns vector with one row per state. Suppose then that the following variables are available within MATLAB®:

Note. Recall that we enumerated the states and actions so that the elements of $\mathcal{X}, \mathcal{U}$, and $\mathcal{D}$ are positive integers.

- F is a cell-array with K elements, each equal to an $n_{\mathcal{X}} \times n_{\mathcal{U}} \times n_{\mathcal{D}}$ three-dimensional matrix so that F{k} represents the game dynamics function $f_k(x, u, d)$, $\forall x \in \mathcal{X}, u \in \mathcal{U}, d \in \mathcal{D}, k \in \{1, 2, \ldots, K\}$. Specifically, the entry F{k}(i,j,l) of the matrix F{k} is the state $f_k(i, j, k)$.
- G is a cell-array with K elements, each equal to an $n_{\mathcal{X}} \times n_{\mathcal{U}} \times n_{\mathcal{D}}$ three-dimensional matrix so that G{k} represents the stage-cost function $g_k(x, u, d)$, $\forall x \in \mathcal{X}, u \in \mathcal{U}, d \in \mathcal{D}, k \in \{1, 2, \ldots, K\}$. Specifically, the entry G{k}(i,j,l) of the matrix G{k} is the per-stage cost $g_k(i, j, k)$.

With these definitions, we can construct $V_k(x)$ in (17.7) very efficiently using the following MATLAB® code:

Note. When this procedure fails because Vminmax and Vmaxmin differ, one may want to use a mixed policy using a linear program. The indices of the states for which this is needed can be found using k=find(Vminmax ~=Vmaxmin).

Note. See Exercise 17.1 for an instantiation of this general code to a specific game. ▷ p. 209

```
V{K+1}=zeros(size(G{K},1),1);
for k=K:-1:1
  Vminmax=min(max(G{k}+V{k+1}(F{k}),[],3),[],2);
  Vmaxmin=max(min(G{k}+V{k+1}(F{k}),[],2),[],3);
  if any(Vminmax~=Vmaxmin)
     error('Saddle-point cannot be found')
     end
  V{k}=Vminmax;
end
```

After running this code, the following variable has been created:

- V is a cell-array with $K + 1$ elements, each equal to an $n_{\mathcal{X}} \times 1$ columns vector so that V{k} represents $V_k(x)$, $\forall x \in \mathcal{X}, k \in \{1, 2, \ldots, K\}$. Specifically, the entry V{k}(i) of the vector V{k} is the cost-to-go $V_k(i)$ from state i at stage k.

For a given state x at stage k, the optimal actions u and d given by (17.9)–(17.10) can be obtained using

```
[~,u]=min(max(G(x,:,:)+V{k+1}(F(x,:,:)),[],3),[],2);
[~,d]=max(min(G(x,:,:)+V{k+1}(F(x,:,:)),[],2),[],3);
```

17.4 Linear Quadratic Dynamic Games

Discrete-time *linear quadratic games* are characterized by linear dynamics of the form

$$x_{k+1} = \underbrace{Ax_k + Bu_k + Ed_k}_{f_k(x_k, u_k, d_k)}, \quad x \in \mathbb{R}^n, \;\; u \in \mathbb{R}^{n_u}, \;\; d \in \mathbb{R}^{n_d}, \;\; k \in \{1, 2, \ldots, K\}$$

and a stage-additive quadratic cost of the form

$$J := \sum_{k=1}^{K}\Big(\|y_k\|^2 + \|u_k\|^2 - \mu^2\|d_k\|^2\Big) = \sum_{k=1}^{K}\underbrace{\Big(x_k'C'Cx_k + u_k'u_k - \mu^2 d_k'd_k\Big)}_{g_k(x_k,u_k,d_k)}$$

where

$$y_k = Cx_k, \quad \forall k \in \{1, 2, \ldots, K\}.$$

This cost function captures scenarios in which:

1. the player P_1 (minimizer) wants to make the $y_k, k \in \{1, 2, \ldots, K\}$ small, without "spending" much effort in their actions $u_k, k \in \{1, 2, \ldots, K\}$;
2. whereas the player P_2 (maximizer) wants to make the same y_k large, without "spending" much effort in their actions $d_k, k \in \{1, 2, \ldots, K\}$.

The constant μ can be seen as a conversion factor that maps units of d_k into units of u_k and y_k.

Note. If needed, a "conversion factor" between units of u and y could be incorporated into the matrix C that defines y.

The equation (17.7) for this game is

$$\begin{aligned}V_k(x) &:= \min_{u_k\in\mathcal{U}_k}\sup_{d_k\in\mathcal{D}_k}\Big(x'C'Cx + u_k'u_k - \mu^2 d_k'd_k + V_{k+1}(Ax + Bu_k + Ed_k)\Big)\\ &= \max_{d_k\in\mathcal{D}_k}\inf_{u_k\in\mathcal{U}_k}\Big(x'C'Cx + u_k'u_k - \mu^2 d_k'd_k + V_{k+1}(Ax + Bu_k + Ed_k)\Big),\end{aligned} \tag{17.11}$$

$\forall x \in \mathbb{R}^n, \ \ k \in \{1, 2, \ldots, K\}$ and, inspired by the quadratic form of the stage cost, we will try to find a solution to (17.11) of the form

$$V_k(x) = x'P_kx, \quad \forall x \in \mathbb{R}^n, \ \ k \in \{1, 2, \ldots, K+1\}$$

for appropriately selected symmetric $n \times n$ matrices P_k. For (17.8) to hold, we need to have $P_{K+1} = 0$ and, for (17.11) to hold, we need

$$\begin{aligned}x'P_kx &= \min_{u_k\in\mathbb{R}^{n_u}}\sup_{d_k\in\mathbb{R}^{n_d}} Q_x(u_k, d_k)\\ &= \max_{d_k\in\mathbb{R}^{n_d}}\inf_{u_k\in\mathbb{R}^{n_u}} Q_x(u_k, d_k), \forall x \in \mathbb{R}^n, \ \ k \in \{1, 2, \ldots, K\},\end{aligned} \tag{17.12}$$

where

$$\begin{aligned}Q_x(u_k, d_k) &:= x'C'Cx + u_k'u_k - \mu^2 d_k'd_k + (Ax + Bu_k + Ed_k)'P_{k+1}(Ax + Bu_k + Ed_k)\\ &= \begin{bmatrix} u_k' & d_k' & x'\end{bmatrix}\begin{bmatrix} I + B'P_{k+1}B & B'P_{k+1}E & B'P_{k+1}A\\ E'P_{k+1}B & -\mu^2 I + E'P_{k+1}E & E'P_{k+1}A\\ A'P_{k+1}B & A'P_{k+1}E & C'C + A'P_{k+1}A\end{bmatrix}\begin{bmatrix} u_k\\ d_k\\ x\end{bmatrix}.\end{aligned}$$

Note. See Exercise 4.3. ▷ p. 50

The right-hand side of (17.12) can be viewed as a quadratic zero-sum game that has a saddle-point equilibrium

$$\begin{bmatrix} u^* \\ d^* \end{bmatrix} = -\begin{bmatrix} I + B'P_{k+1}B & B'P_{k+1}E \\ E'P_{k+1}B & -\mu^2 I + E'P_{k+1}E \end{bmatrix}^{-1} \begin{bmatrix} B'P_{k+1}A \\ E'P_{k+1}A \end{bmatrix} x,$$

with value given by

$$x'\Bigg(C'C + A'P_{k+1}A - [\, A'P_{k+1}B \quad A'P_{k+1}E \,] \begin{bmatrix} I + B'P_{k+1}B & B'P_{k+1}E \\ E'P_{k+1}B & -\mu^2 I + E'P_{k+1}E \end{bmatrix}^{-1} \begin{bmatrix} B'P_{k+1}A \\ E'P_{k+1}A \end{bmatrix}\Bigg) x,$$

provided that

$$I + B'P_{k+1}B > 0, \qquad -\mu^2 I + E'P_{k+1}E < 0.$$

In this case, (17.12) holds provided that

$$P_k = C'C + A'P_{k+1}A - [\, A'P_{k+1}B \quad A'P_{k+1}E \,] \begin{bmatrix} I + B'P_{k+1}B & B'P_{k+1}E \\ E'P_{k+1}B & -\mu^2 I + E'P_{k+1}E \end{bmatrix}^{-1} \begin{bmatrix} B'P_{k+1}A \\ E'P_{k+1}A \end{bmatrix}.$$

In this case, Theorem 17.1 can be used to compute saddle-point equilibria for this game and leads to the following result.

Corollary 17.1. Suppose that we define the matrices P_k according to the following (backwards) recursion:

$$P_{K+1} = 0$$

$$P_k = C'C + A'P_{k+1}A - [\, A'P_{k+1}B \quad A'P_{k+1}E \,] \begin{bmatrix} I + B'P_{k+1}B & B'P_{k+1}E \\ E'P_{k+1}B & -\mu^2 I + E'P_{k+1}E \end{bmatrix}^{-1} \begin{bmatrix} B'P_{k+1}A \\ E'P_{k+1}A \end{bmatrix},$$

$\forall k \in \{1, 2, \ldots, K\}$ and that

$$I + B'P_{k+1}B > 0, \quad -\mu^2 I + E'P_{k+1}E < 0, \quad \forall k \in \{1, 2, \ldots, K\}.$$

Then the pair of policies (γ^*, σ^*) defined as follows is a saddle-point equilibrium in state-feedback policies:

$$\begin{bmatrix} \gamma_k^*(x) \\ \sigma_k^*(x) \end{bmatrix} = -\begin{bmatrix} I + B'P_{k+1}B & B'P_{k+1}E \\ E'P_{k+1}B & -\mu^2 I + E'P_{k+1}E \end{bmatrix}^{-1} \begin{bmatrix} B'P_{k+1}A \\ E'P_{k+1}A \end{bmatrix} x,$$

$\forall x \in \mathcal{X}$, $k \in \{1, 2, \ldots, K\}$. Moreover, the value of the game is equal to $x_1' P_1 x_1$. □

Note (Induced norm). Since (γ^*, σ^*) is a saddle-point equilibrium with value $x_1'P_1x_1$, when P_1 uses their security policy

$$u_k = \gamma_k^*(x_k), \tag{17.13}$$

for every policy $d_k = \sigma_k(x_k)$ for P_2, we have that

$$J(\gamma^*, \sigma^*) = x_1' P_1 x_1 \geq J(\gamma^*, \sigma) = \sum_{k=1}^{K} \Big(\|y_k\|^2 + \|u_k\|^2 - \mu^2 \|d_k\|^2 \Big)$$

and therefore

$$\sum_{k=1}^{K} \|y_k\|^2 \leq x_1' P_1 x_1 + \mu^2 \sum_{k=1}^{K} \|d_k\|^2 - \sum_{k=1}^{K} \|u_k\|^2.$$

When $x_1 = 0$, this implies that

$$\sum_{k=1}^{K} \|y_k\|^2 \leq \mu^2 \sum_{k=1}^{K} \|d_k\|^2.$$

Moreover, in view of Note 16 (p. 205), this holds for every possible d_k, regardless of the information structure available to P_2, and therefore we conclude that

$$\sup_{d_k,\ k \in \{1, \ldots, K\}} \frac{\sqrt{\sum_{k=1}^{K} \|y_k\|^2}}{\sqrt{\sum_{k=1}^{K} \|d_k\|^2}} \leq \mu. \tag{17.14}$$

Notation. When $K = \infty$, the left-hand side of (18.8) is called the discrete-time H-infinity norm of the closed-loop and (18.7) guarantees an H-infinity norm smaller than or equal to μ.

In view of (17.14), the control law (17.13) is said to achieve an *$\mathcal{L}_2$-induced norm from the disturbance d_k, $k \in \{1, 2, \ldots, K\}$ to the output y_k, $k \in \{1, 2, \ldots, K\}$ lower than or equal to μ.*

17.5 Practice Exercise

17.1 (Tic-Tac-Toe). Write a MATLAB® script to compute the cost-to-go for each state of the Tic-Tac-Toe game. Assume that player P_1 (minimizer) places the Xs and player P_2 (maximizer) places the Os. The game outcome should be -1 when P_1 wins, $+1$ when P_2 wins, and 0 when the game ends in a draw.

Hint: Draw inspiration from the code in Section 17.3, but keep in mind that Tic-Tac-Toe is a game of alternate play, whereas the general algorithm in Section 17.3 is for simultaneous play.

Solution to Exercise 17.1. We start by discussing the choices made for the design of the MATLAB® code and then provide the final code.

Alternate play:. A simple approach to convert an alternate-play game like Tic-Tac-Toe into a simultaneous-play game like the ones we have addressed in this section, consists of expanding each stage of the alternate-play game into 2 sequential stages of a simultaneous-play game. For our Tic-Tac-Toe game, in stage 1 player P_1 selects where to place the first X, but player P_2 cannot place any O. Then, in stage 2, P_2 selects where to place an O, but P_1 cannot place any X. This continues, with P_1 placing Xs in stages 1, 3, 5, 7, and 9 and P_2 placing Os in stages 2, 4, 6, and 8. In this expanded 9-stage game, one can imagine that, at each stage, both players play simultaneously, but, in practice, one of the players has no choice to make.

State encoding:. We encode the possible states of the game by assigning to each state an 18-bit integer. Each pair of bits in this integer is associated with one of the 9 slots in the Tic-Tac-Toe board as follows:

Bit #	17	16	15	14	13	12	11	10	9	8	7	6	5	4	3	2	1	0
Slot	1		2		3		4		5		6		7		8		9	

where the 9 slots are numbered as follows:

1	2	3
4	5	6
7	8	9

The two bits associated with a slot indicate its content:

Most significant bit	Least significant bit	Meaning
0	0	empty slot
0	1	X
1	0	O
1	1	invalid

With this encoding, the MATLAB® function `ttt_addX(Sk)` defined below takes an $N \times 1$ vector `Sk` of integers representing states and generates an $N \times 9$ matrix `newS` that, for each of the N states in `Sk`, computes all the states that would be obtained by adding an X to each of the 9 possible slots. The same function generates two additional outputs:

Note. For efficiency, the code that computes `newS`, `invalid`, and `won` is vectorized in that it does not loop over the N states in `Sk`. This enables very fast computations, even for a very large number N of states in `Sk`.

- an $N \times 9$ Boolean-valued matrix `invalid` where an entry equal to `true` indicates that the corresponding entry in `newS` does not correspond to a valid placement of an X because the corresponding slot was not empty; and
- an $N \times 9$ Boolean-valued matrix `won` where an entry equal to `true` indicates that the corresponding entry in `newS` has three Xs in a row.

```
function [newS,won,invalid]=ttt_addX(Sk)
    XplayMasks=int32([bin2dec('010000 000000 000000');
                      bin2dec('000100 000000 000000');
                      bin2dec('000001 000000 000000');
                      bin2dec('000000 010000 000000');
                      bin2dec('000000 000100 000000');
                      bin2dec('000000 000001 000000');
                      bin2dec('000000 000000 010000');
                      bin2dec('000000 000000 000100');
                      bin2dec('000000 000000 000001')]);
    % compute new state and test whether move is valid
    newS=zeros(size(Sk,1),length(XplayMasks),'int32');
    invalid=false(size(newS));
    for slot=1:length(XplayMasks)
        mask=XplayMasks(slot);
```

```
        newS(:,slot)=bitor(S,mask);
        invalid(bitand(Sk,mask+2*mask)~=0,slot)=true;
    end

    XwinMasks=int32([bin2dec('010101 000000 000000');   % top horizontal
                     bin2dec('000000 010101 000000');   % mid horizontal
                     bin2dec('000000 000000 010101');   % bottom horizontal
                     bin2dec('010000 010000 010000');   % left vertical
                     bin2dec('000100 000100 000100');   % center vertical
                     bin2dec('000001 000001 000001');   % right vertical
                     bin2dec('010000 000100 000001');   % descend diagonal
                     bin2dec('000001 000100 010000')]); % ascend diagonal
    % check if X won
    won=false(size(newS));
    for i=1:length(XwinMasks)
        won=bitor(won,bitand(newS,XwinMasks(i))==XwinMasks(i));
    end
end
```

The function `ttt_addO(Sk)` plays a similar role, but now for adding Os.

```
function [newS,won,invalid]=ttt_addO(Sk,slot)
    OplayMasks=int32([bin2dec('100000 000000 000000');
                      bin2dec('001000 000000 000000');
                      bin2dec('000010 000000 000000');
                      bin2dec('000000 100000 000000');
                      bin2dec('000000 001000 000000');
                      bin2dec('000000 000010 000000');
                      bin2dec('000000 000000 100000');
                      bin2dec('000000 000000 001000');
                      bin2dec('000000 000000 000010')]);
    % compute new state and test whether move is valid
    newS=zeros(size(Sk,1),length(OplayMasks),'int32');
    invalid=false(size(newS));
    for slot=1:length(OplayMasks)
        mask=OplayMasks(slot);
        newS(:,slot)=bitor(Sk,mask);
        invalid(bitand(Sk,mask+mask/2)~=0,slot)=true;
    end

    OwinMasks=int32([bin2dec('101010 000000 000000');   % top horizontal
                     bin2dec('000000 101010 000000');   % mid horizontal
                     bin2dec('000000 000000 101010');   % bottom horizontal
                     bin2dec('100000 100000 100000');   % left vertical
                     bin2dec('001000 001000 001000');   % center vertical
                     bin2dec('000010 000010 000010');   % right vertical
                     bin2dec('100000 001000 000010');   % descend diagonal
                     bin2dec('000010 001000 100000')]); % ascend diagonal
    % check if O won
    won=false(size(newS));
    for i=1:length(OwinMasks)
        won=bitor(won,bitand(newS,OwinMasks(i))==OwinMasks(i));
    end
end
```

State enumeration:. To compute the cost-to-go, we need to enumerate all the states that can occur at each stage of the Tic-Tac-Toe game. This can be accomplished by the following function.

```
function S=ttt_states(S0)
    K=9;
    S=cell(K+1,1);
    S{1}=S0;
    for k=1:K
        if rem(k,2)==1
            % player X (minimizer) plays at odd stages
            [newS,won,invalid]=ttt_addX(S{k}); % compute all next states
        else
            % player O (minimizer) plays at even stages
            [newS,won,invalid]=ttt_addO(S{k}); % compute all next states
        end
        % stack all states in a column vector
        newS=reshape(newS,[],1);
        won=reshape(won,[],1);
        invalid=reshape(invalid,[],1);
        % store (unique) list of states for which the game continues
        S{k+1}=unique(newS(~invalid & ~won));
    end
end
```

This function returns a cell-array S with 10 elements. Each entry S{k} of S is a vector containing all valid stage-k states for which the game has not yet finished. We remove from S{k} all the "game-over" states because we do not need to compute the cost-to-go for such states.

Final code. The following code computes the cost-to-go for each state in the cell-array S computed by the function `ttt_states()`.

```
K=9;
V=cell(K+1,1);
V{K+1}=zeros(size(S{K+1}),'int8');
for k=K:-1:1
    if rem(k,2)==1
        % player X (minimizer) plays at odd stages
        [newS,won,invalid]=ttt_addX(S{k}); % compute all next states
        % convert states to indices in S{k+1}
        % to get their costs-to-go from V{k+1}
        [exists,newSndx]=ismember(newS,S{k+1});
        % compute all possible values
        newV=zeros(size(newS),'int8');
        newV(exists)=V{k+1}(newSndx(exists));
        newV(won)=-1;
        newV(invalid)=+Inf; % penalize invalid actions for minimizer
        V{k}=min(newV,[],2); % pick best for minimizer
    else
        % player O (maximizer) plays at even stages
        [newS,won,invalid]=ttt_addO(S{k}); % compute all next states
        % convert states to indices in S{k+1}
```

```
            % to get their costs-to-go from V{k+1}
            [exists,newS]=ismember(newS,S{k+1});
            % compute all possible values
            newV=zeros(size(newS),'int8');
            newV(exists)=V{k+1}(newS(exists));
            newV(won)=1;
            newV(invalid)=-Inf; % penalize invalid actions for maximizer
            V{k}=max(newV,[],2);  % pick best for maximizer
        end
    end
```

This code returns a cell-array V with 10 elements. Each entry `V{k}` of V is an array with the same size as `S{k}` whose entries are equal to the cost-to-go from the corresponding state in `S{k}` at stage `k`.

The code above has the same basic structure as the general code in Section 17.3, but it is optimized to take advantage of the specific structure of this game:

- Since P_1 places an X at the odd stages and P_2 places an O the even stages, we find an `if` statement inside the `for` loop that allows the construction of the cost-to-go `V{k}` to differ depending on whether `k` is even or odd.
- For the code in Section 17.3, the matrix `F{k}` contains all possible states that can be reached at stage `k+1` for all possible actions for each player. Here, the functions `ttt_addX(S{k})` and `ttt_addO(S{k})` provide this set of states at the even and odd stages, respectively. In essence, the variable `newS` corresponds to `F{k}` in the code in Section 17.3, with the caveat that `newS` contains some invalid states that need to be ignored.
- The code in Section 17.3 uses `G{k}+V{k+1}(F{k})` to add the per-stage cost `G{k}` at stage `k` with the cost-to-go `V{k+1}(F{k})` from stage `k+1`. In our Tic-Tac-Toe game, the per-stage cost is always zero unless the games finishes, so there is no need to add the per-stage cost until one of the players wins. Moreover, when a player wins, we do not need to consider the cost-to-go from subsequent stages because the game will end. In essence, the variable `newV` corresponds to `G{k}+V{k+1}(F{k})` in the code in Section 17.3.
- When `k` is odd, only P_1 (minimizer) can make a choice so there is no maximization to carry out over the actions of P_2. This means that `Vminmax` and `Vmaxmin` are both obtained with a simple minimization and are always equal to each other. Alternatively, when `k` is even only P_2 (maximizer) can make a choice, which means that `Vminmax` and `Vmaxmin` are both obtained with a simple maximization and are always equal to each other. This means that here we do not need to compute `Vminmax` and `Vmaxmin` and test if they are equal, before assigning their value to `V{k}`.

LECTURE 18

State-Feedback Zero-Sum Differential Games

This lecture addresses the computation of saddle-point equilibria of zero-sum continuous-time dynamic games in state-feedback policies.

18.1 Zero-Sum Dynamic Games in Continuous Time

We now discuss the solution for two-player zero-sum dynamic games in continuous time, which corresponds to dynamics of the form

$$\underbrace{\dot{x}(t)}_{\substack{\text{state}\\ \text{derivative}}} = \underbrace{f}_{\substack{\text{game}\\ \text{dynamics}}}\Big(\underbrace{t}_{\text{time}}, \underbrace{x(t)}_{\substack{\text{current}\\ \text{state}}}, \underbrace{u(t)}_{\substack{\mathsf{P}_1\text{'s action}\\ \text{at time } t}}, \underbrace{d(t)}_{\substack{\mathsf{P}_2\text{'s action}\\ \text{at time } t}}\Big), \quad \forall t \in [0, T],$$

with state $x(t) \in \mathbb{R}^n$ initialized at a given $x(0) = x_0$. For every time $t \in [0, T]$, P_1's action $u(t)$ is required to belong to an action space $\mathcal{U}$ and P_2's action $d(t)$ is required to belong to an action space $\mathcal{D}$. We assume a finite horizon ($T < \infty$) integral cost of the form

Note. For simplicity, we assume here that the action spaces are the same at all times, but it would be straightforward to lift that restriction.

$$J := \int_0^T g(t, x(t), u(t), d(t))dt + q(x(T))$$

that P_1 wants to minimize and P_2 wants to maximize. In this section we consider a *state-feedback information structure*, which corresponds to policies of the form

$$u(t) = \gamma(t, x(t)), \quad d(t) = \sigma(t, x(t)), \quad \forall t \in [0, T].$$

For continuous-time we can also use dynamic programming to construct saddle-point equilibria in state-feedback policies. The following result is the equivalent of Theorem 17.1 for continuous time.

Notation. By using a max and min (instead of sup and inf) in the outer optimizations in (18.1), we implicitly require the outer sup and inf to be achieved at some point in the sets.

Theorem 18.1. Assume that there exists a continuously differentiable function $V(t, x)$ that satisfies the following *Hamilton-Jacobi-Bellman-Isaac* equation

$$-\frac{\partial V(t,x)}{\partial t} = \min_{u \in \mathcal{U}} \sup_{d \in \mathcal{D}} \left(g(t,x,u,d) + \frac{\partial V(t,x)}{\partial x} f(t,x,u,d) \right) \tag{18.1a}$$

$$= \max_{d \in \mathcal{D}} \inf_{u \in \mathcal{U}} \left(g(t,x,u,d) + \frac{\partial V(t,x)}{\partial x} f(t,x,u,d) \right), \quad \forall x \in \mathbb{R}^n, \quad t \in [0, T] \tag{18.1b}$$

Note. We can view (18.2) as a boundary condition for the Hamilton-Jacobi-Bellman-Isaac equation (18.1), which is a partial differential equation (PDE).

with

$$V(T, x) = q(x), \quad \forall x \in \mathbb{R}^n. \tag{18.2}$$

Then the pair of policies (γ^*, σ^*) defined as follows is a saddle-point equilibrium in state-feedback policies:

$$\gamma^*(t,x) := \arg\min_{u \in \mathcal{U}} \sup_{d \in \mathcal{D}} \left(g(t,x,u,d) + \frac{\partial V(t,x)}{\partial x} f(t,x,u,d) \right)$$

$$\sigma^*(t,x) := \arg\max_{d \in \mathcal{D}} \inf_{u \in \mathcal{U}} \left(g(t,x,u,d) + \frac{\partial V(t,x)}{\partial x} f(t,x,u,d) \right),$$

$\forall x \in \mathbb{R}^n, \quad t \in [0, T]$. Moreover, the value of the game is equal to $V(0, x_0)$. □

Attention! Theorem 18.1 provides a *sufficient condition* for the existence of Nash equilibria, but this condition is not necessary. In particular, the two security levels in (18.1) may not commute for some state x at some stage t, but there still may be a saddle-point for the game.

Proof of Theorem 18.1. From the fact that the inf and sup commute in (18.1) and the definitions of $\gamma^*(t, x)$ and $\sigma^*(t, x)$, we conclude that the pair $(\gamma^*(t, x), \sigma^*(t, x))$ is a saddle-point equilibrium for a zero-sum game with criterion

$$g(t,x,u,d) + \frac{\partial V(t,x)}{\partial x} f(t,x,u,d),$$

which means that

$$\begin{aligned}
&g(t,x,\gamma^*(t,x),d) + \frac{\partial V(t,x)}{\partial x} f(t,x,\gamma^*(t,x),d) \\
&\quad \le g(t,x,\gamma^*(t,x),\sigma^*(t,x)) + \frac{\partial V(t,x)}{\partial x} f(t,x,\gamma^*(t,x),\sigma^*(t,x)) \\
&\quad \le g(t,x,u,\sigma^*(t,x)) + \frac{\partial V(t,x)}{\partial x} f(t,x,u,\sigma^*(t,x)), \quad \forall u \in \mathcal{U}, \ d \in \mathcal{D}.
\end{aligned}$$

Note 17. We can actually conclude that P_2 cannot get a reward larger than $V(0, x_0)$ against $\gamma^*(t, x)$, regardless of the information structure available to P_2. ▷ p. 216

Moreover, since the middle term in these inequalities is also equal to the right-hand-side of (18.1), we have that

$$-\frac{\partial V(t,x)}{\partial t} = g(t, x, \gamma^*(t,x), \sigma^*(t,x)) + \frac{\partial V(t,x)}{\partial x} f(t, x, \gamma^*(t,x), \sigma^*(t,x))$$
$$= \sup_{d \in \mathcal{D}} \Big(g(t, x, \gamma^*(t,x), d) + \frac{\partial V(t,x)}{\partial x} f(t, x, \gamma^*(t,x), d) \Big),$$
$$\forall x \in \mathbb{R}^n, \ t \in [0, T]$$

which, because of Theorem 16.1, shows that $\sigma^*(t, x)$ is an optimal (maximizing) state-feedback policy against $\gamma^*(t, x)$ and the maximum is equal to $V(0, x_0)$. Moreover, since we also have that

Note 17. We can actually conclude that P_1 cannot get a cost smaller than $V(0, x_0)$ against $\sigma^*(t, x)$, regardless of the information structure available to P_1. ▷ p. 216

$$-\frac{\partial V(t,x)}{\partial t} = g(t, x, \gamma^*(t,x), \sigma^*(t,x)) + \frac{\partial V(t,x)}{\partial x} f(t, x, \gamma^*(t,x), \sigma^*(t,x))$$
$$= \inf_{u \in \mathcal{U}} \left(g(t, x, u, \sigma^*(t,x)) + \frac{\partial V(t,x)}{\partial x} f(t, x, u, \sigma^*(t,x)) \right),$$
$$\forall x \in \mathbb{R}^n, \ t \in [0, T],$$

we can also conclude that $\gamma^*(t, x)$ is an optimal (minimizing) state-feedback policy against $\sigma^*(t, x)$ and the minimum is also equal to $V(0, x_0)$. This proves that (γ^*, σ^*) is indeed a saddle-point equilibrium in state-feedback policies with value $V(0, x_0)$. ■

Note 17. In view of Note 15 (p. 194), we actually conclude that:

1. P_2 cannot get a reward larger than $V(0, x_0)$ against $\gamma^*(t, x)$, regardless of the information structure available to P_2, and
2. P_1 cannot get a cost smaller than $V(0, x_0)$ against $\sigma^*(t, x)$, regardless of the information structure available to P_1.

In practice, this means that $\gamma^*(t, x)$ and $\sigma^*(t, x)$ are "extremely safe" policies for P_1 and P_2, respectively, since they guarantee a level of reward regardless of the information structure for the other player. □

18.2 Linear Quadratic Dynamic Games

Continuous-time *linear quadratic games* are characterized by linear dynamics of the form

$$\dot{x} = \underbrace{Ax(t) + Bu(t) + Ed(t)}_{f(t,x(t),u(t),d(t))}, \quad x \in \mathbb{R}^n, \ u \in \mathbb{R}^{n_u}, \ d \in \mathbb{R}^{n_d}, \ t \in [0, T]$$

and an integral quadratic cost of the form

$$J := \int_0^T \underbrace{\left(\|y(t)\|^2 + \|u(t)\|^2 - \mu^2 \|d(t)\|^2 \right)}_{g(t,x(t),u(t),d(t))} dt + \underbrace{x'(T) P_T x(T)}_{q(x(T))},$$

where

$$y(t) = Cx(t), \quad \forall t \in [0, T].$$

This cost function captures scenarios in which:

1. the player P_1 (minimizer) wants to make $y(t)$ small over the interval $[0, T]$ without "spending" much effort in their action $u(t)$,
2. whereas the player P_2 (maximizer) wants to make $y(t)$ large without "spending" much effort in their action $d(t)$.

The constant μ can be seen as a conversion factor that maps units of $d(t)$ into units of $u(t)$ and $y(t)$.

Note. If needed, a "conversion factor" between units of u and y could be incorporated into the matrix C that defines y.

The Hamilton-Jacobi-Bellman-Isaac equation (18.1) for this game is

$$-\frac{\partial V(t,x)}{\partial t} = \min_{u\in\mathbb{R}^{n_u}} \sup_{d\in\mathbb{R}^{n_d}} Q(u,d) = \max_{d\in\mathbb{R}^{n_d}} \inf_{u\in\mathbb{R}^{n_u}} Q(u,d), \quad \forall x \in \mathbb{R}^n,\ t\in[0,T], \tag{18.3}$$

where

$$Q(u,d) := x'C'Cx + u'u - \mu^2 d'd + \frac{\partial V(t,x)}{\partial x}(Ax + Bu + Ed)$$
$$= \begin{bmatrix} u' & d' & x' & \frac{\partial V(t,x)}{\partial x} \end{bmatrix} \begin{bmatrix} I & 0 & 0 & \frac{1}{2}B' \\ 0 & -\mu^2 I & 0 & \frac{1}{2}E' \\ 0 & 0 & C'C & \frac{1}{2}A' \\ \frac{1}{2}B & \frac{1}{2}E & \frac{1}{2}A & 0 \end{bmatrix} \begin{bmatrix} u \\ d \\ x \\ \frac{\partial V(t,x)}{\partial x}' \end{bmatrix}.$$

Note. See Exercise 4.3. ▷ p. 50

The right-hand side of (18.3) can be viewed as a quadratic zero-sum game that has a saddle-point equilibrium

$$\begin{bmatrix} u^* \\ d^* \end{bmatrix} = -\begin{bmatrix} I & 0 \\ 0 & -\mu^2 I \end{bmatrix}^{-1} \begin{bmatrix} 0 & \frac{1}{2}B' \\ 0 & \frac{1}{2}E' \end{bmatrix} \begin{bmatrix} x \\ \frac{\partial V(t,x)}{\partial x}' \end{bmatrix} = \frac{1}{2}\begin{bmatrix} -B' \\ \mu^{-2}E' \end{bmatrix} \frac{\partial V(t,x)}{\partial x}',$$

with value given by

$$\begin{bmatrix} x' & \frac{\partial V(t,x)}{\partial x} \end{bmatrix} \left(\begin{bmatrix} C'C & \frac{1}{2}A' \\ \frac{1}{2}A & 0 \end{bmatrix} - \begin{bmatrix} 0 & 0 \\ \frac{1}{2}B & \frac{1}{2}E \end{bmatrix} \begin{bmatrix} I & 0 \\ 0 & -\mu^2 I \end{bmatrix}^{-1} \begin{bmatrix} 0 & \frac{1}{2}B' \\ 0 & \frac{1}{2}E' \end{bmatrix} \right) \begin{bmatrix} x \\ \frac{\partial V(t,x)}{\partial x}' \end{bmatrix}$$
$$= \begin{bmatrix} x' & \frac{\partial V(t,x)}{\partial x} \end{bmatrix} \begin{bmatrix} C'C & \frac{1}{2}A' \\ \frac{1}{2}A & \frac{1}{4}\mu^{-2}EE' - \frac{1}{4}BB' \end{bmatrix} \begin{bmatrix} x \\ \frac{\partial V(t,x)}{\partial x}' \end{bmatrix}.$$

This means that this game's Hamilton-Jacobi-Bellman-Isaac equation (18.3) is a PDE of the form

$$-\frac{\partial V(t,x)}{\partial t} = \begin{bmatrix} x' & \frac{\partial V(t,x)}{\partial x} \end{bmatrix} \begin{bmatrix} C'C & \frac{1}{2}A' \\ \frac{1}{2}A & \frac{1}{4}\mu^{-2}EE' - \frac{1}{4}BB' \end{bmatrix} \begin{bmatrix} x \\ \frac{\partial V(t,x)}{\partial x}' \end{bmatrix}, \tag{18.4}$$

$\forall x \in \mathbb{R}^n, \ t \in [0, T]$. Inspired by the boundary condition

$$V(T, x) = x' P_T x, \quad \forall x \in \mathbb{R}^n,$$

we will try to find a solution to (18.4) of the form

$$V(t, x) = x' P(t) x, \quad \forall x \in \mathbb{R}^n, \ t \in [0, T]$$

for some appropriately selected symmetric $n \times n$ matrix $P(t)$. For this function, (18.4) becomes

$$-x'\dot{P}(t)x = [\, x' \quad 2x'P(t)\,] \begin{bmatrix} C'C & \frac{1}{2}A' \\ \frac{1}{2}A & \frac{1}{4}\mu^{-2}EE' - \frac{1}{4}BB' \end{bmatrix} \begin{bmatrix} x \\ 2P(t)x \end{bmatrix}$$

$$= x'\Big(C'C + P(t)A + A'P(t) + P(t)(\mu^{-2}EE' - BB')P(t)\Big)x, \quad (18.5)$$

$\forall x \in \mathbb{R}^n, \ t \in [0, T]$, which holds provided that

$$-\dot{P}(t) = C'C + P(t)A + A'P(t) + P(t)(\mu^{-2}EE' - BB')P(t), \quad \forall t \in [0, T].$$

In this case, Theorem 18.1 can be used to compute saddle-point equilibria for this game and leads to the following result.

Note. The function $P(t)$ could be found by numerically solving the matrix-valued ordinary differential equation (18.6) backwards in time.

Corollary 18.1. Suppose that there is a symmetric solution to the following matrix-valued ordinary differential equation

$$-\dot{P}(t) = C'C + P(t)A + A'P(t) + P(t)(\mu^{-2}EE' - BB')P(t), \quad \forall t \in [0, T], \quad (18.6)$$

with final condition $P(T) = P_T$. Then the state-feedback policies

$$\gamma^*(t, x) = -B'P(t)x, \quad \sigma^*(t, x) = \mu^{-2}E'P(t)x, \quad \forall x \in \mathbb{R}^n, \ t \in [0, T]$$

form a saddle-point equilibrium in state-feedback policies with value $x_0'P(0)x_0$. □

Note (Induced norm). Since (γ^*, σ^*) is a saddle-point equilibrium with value $x_0'P(0)x_0$, when P_1 uses its security policy

$$u(t) = \gamma^*(t, x(t)) = -B'Px(t), \quad (18.7)$$

for every policy $d(t) = \sigma(t, x(t))$ for P_2, we have that

$$J(\gamma^*, \sigma^*) = x_0'P(0)x_0 \geq J(\gamma^*, \sigma)$$

$$= \int_0^T \Big(\|y(t)\|^2 + \|u(t)\|^2 - \mu^2\|d(t)\|^2\Big)dt + x'(T)P_T x(T)$$

and therefore

$$\int_0^T \|y(t)\|^2 dt \leq x_0'P(0)x_0 + \mu^2 \int_0^T \|d(t)\|^2 dt - \int_0^T \|u(t)\|^2 dt - x'(T)P_T x(T).$$

When P_T is positive semi-definite and $x_0 = 0$, this implies that

$$\int_0^T \|y(t)\|^2 dt \le \mu^2 \int_0^T \|d(t)\|^2 dt.$$

Moreover, in view of Note 17 (p. 216), this holds for every possible $d(t)$, regardless of the information structure available to P_2, and therefore we conclude that

Notation. When $T = \infty$, the left-hand side of (18.8) is called the H-infinity norm of the closed-loop and (18.7) guarantees an H-infinity norm lower than or equal to μ.

$$\sup_{d(t),\, t \in [0,T]} \frac{\sqrt{\int_0^T \|y(t)\|^2 dt}}{\sqrt{\int_0^T \|d(t)\|^2 dt}} \le \mu. \tag{18.8}$$

In view of (18.8), the control law (18.7) is said to achieve an *$\mathcal{L}_2$-induced norm in the interval* $[0, T]$ *from the disturbance d to the output y lower than or equal to* μ.

18.3 Differential Games with Variable Termination Time

Consider now a two-player zero-sum differential game with the usual dynamics

$$\dot{x}(t) = f(x(t), u(t), d(t)), \quad x(t) \in \mathbb{R}^n, \; u(t) \in \mathcal{U}, \; d(t) \in \mathcal{D}, \; t \ge 0,$$

and initialized at a given $x(0) = x_0$, but with an integral cost with variable horizon.

$$J := \int_0^{T_{\text{end}}} g(t, x(t), u(t), d(t))dt + q(T_{\text{end}}, x(T_{\text{end}})),$$

where T_{end} is the first time at which $x(t)$ enters a closed set $\mathcal{X}_{\text{end}} \subset \mathbb{R}^n$ or $T_{\text{end}} = +\infty$ in case $x(t)$ never enters $\mathcal{X}_{\text{end}}$.

Notation. We can think of $\mathcal{X}_{\text{end}}$ as the set of states at which the game terminates, as the evolution of $x(t)$ is irrelevant after this time. The states in $\mathcal{X}_{\text{end}}$ are often called the *game-over* states.

Also for this game we can use dynamic programming to construct saddle-point equilibria in state-feedback policies. The following result is the equivalent of Theorem 18.1 for this game with variable termination time.

Theorem 18.2. Assume that there exists a continuously differentiable function $V(t, x)$ that satisfies the *Hamilton-Jacobi-Bellman-Isaac* equation (18.1) with

Note. We can view (18.9) as a boundary condition for the Hamilton-Jacobi-Bellman-Isaac equation (18.1). From that perspective, Theorems 18.1 and 18.2 share the same Hamilton-Jacobi-Bellman-Isaac PDE and only differ by the boundary conditions.

$$V(t, x) = q(t, x), \quad \forall t \ge 0, \; x \in \mathcal{X}_{\text{end}}. \tag{18.9}$$

Then the pair of policies (γ^*, σ^*) defined as follows is a saddle-point equilibrium in state-feedback policies:

$$\gamma^*(t, x) := \arg\min_{u \in \mathcal{U}} \sup_{d \in \mathcal{D}} \Big(g(t, x, u, d) + \frac{\partial V(t, x)}{\partial x} f(t, x, u, d) \Big)$$

$$\sigma^*(t, x) := \arg\max{}_{d \in \mathcal{D}} \inf_{u \in \mathcal{U}} \Big(g(t, x, u, d) + \frac{\partial V(t, x)}{\partial x} f(t, x, u, d) \Big),$$

$\forall x \in \mathbb{R}^n$, $t \in [0, T]$. Moreover, the value of the game is equal to $V(0, x_0)$. □

Proof of Theorem 18.2. From the fact that the inf and sup commute in (18.1) and the definitions of $\gamma^*(t,x)$ and $\sigma^*(t,x)$, we have that

$$
\begin{aligned}
g(t,x,\gamma^*(t,x),d) &+ \frac{\partial V(t,x)}{\partial x} f(t,x,\gamma^*(t,x),d) \\
&\leq g(t,x,\gamma^*(t,x),\sigma^*(t,x)) + \frac{\partial V(t,x)}{\partial x} f(t,x,\gamma^*(t,x),\sigma^*(t,x)) \\
&\leq g(t,x,u,\sigma^*(t,x)) + \frac{\partial V(t,x)}{\partial x} f(t,x,u,\sigma^*(t,x)), \quad \forall u \in \mathcal{U}, \ d \in \mathcal{D}.
\end{aligned}
$$

Note 18. We can actually conclude that P_2 cannot get a reward larger than $V(0,x_0)$ against $\gamma^*(t,x)$, regardless of the information structure available to P_2. ▷ p. 220

Moreover, since the middle term in these inequalities is also equal to the right-hand-side of (18.1), we have that

$$
\begin{aligned}
-\frac{\partial V(t,x)}{\partial t} &= g(t,x,\gamma^*(t,x),\sigma^*(t,x)) + \frac{\partial V(t,x)}{\partial x} f(t,x,\gamma^*(t,x),\sigma^*(t,x)) \\
&= \sup_{d \in \mathcal{D}} \Big(g(t,x,\gamma^*(t,x),d) + \frac{\partial V(t,x)}{\partial x} f(t,x,\gamma^*(t,x),d) \Big), \\
&\quad \forall x \in \mathbb{R}^n, \quad t \in [0,T]
\end{aligned}
$$

which, because of Theorem 16.2, shows that $\sigma^*(t,x)$ is an optimal (maximizing) state-feedback policy against $\gamma^*(t,x)$ and the maximum is equal to $V(0,x_0)$. Moreover, since we also have that

Note 18. We can actually conclude that P_1 cannot get a cost smaller than $V(0,x_0)$ against $\sigma^*(t,x)$, regardless of the information structure available to P_1. ▷ p. 220

$$
\begin{aligned}
-\frac{\partial V(t,x)}{\partial t} &= g(t,x,\gamma^*(t,x),\sigma^*(t,x)) + \frac{\partial V(t,x)}{\partial x} f(t,x,\gamma^*(t,x),\sigma^*(t,x)) \\
&= \inf_{u \in \mathcal{U}} \Big(g(t,x,u,\sigma^*(t,x)) + \frac{\partial V(t,x)}{\partial x} f(t,x,u,\sigma^*(t,x)) \Big), \\
&\quad \forall x \in \mathbb{R}^n, \quad t \in [0,T]
\end{aligned}
$$

we can also conclude that $\gamma^*(t,x)$ is an optimal (minimizing) state-feedback policy against $\sigma^*(t,x)$ and the minimum is also equal to $V(0,x_0)$. This proves that (γ^*,σ^*) is indeed a saddle-point equilibrium in state-feedback policies with value $V(0,x_0)$. ■

Note 18. In view of Note 15 (p. 194), we actually conclude that:

1. P_2 cannot get a reward larger than $V(0,x_0)$ against $\gamma^*(t,x)$, regardless of the information structure available to P_2, and
2. P_1 cannot get a cost smaller than $V(0,x_0)$ against $\sigma^*(t,x)$, regardless of the information structure available to P_1. □

18.4 Pursuit-Evasion

Pursuit-evasion games are characterized by a state vector $x(t)$ with two distinct vector-valued components $x_1(t)$ and $x_2(t)$, one associated to each player. The dynamics of the players are typically decoupled and of the following form

$$\begin{cases} \dot{x}_1(t) = f_1(x_1, u) & x_1(t) \in \mathbb{R}^{n_1}, u(t) \in \mathcal{U} \\ \dot{x}_2(t) = f_2(x_2, d) & x_2(t) \in \mathbb{R}^{n_2}, d(t) \in \mathcal{D}, \end{cases} \quad \forall t \geq 0.$$

The outcome of such games is determined by a (closed) *capture set* $\mathcal{X}_{\text{end}} \subset \mathbb{R}^{n_1} \times R^{n_2}$. We say that the pursuer P_1 *captures* the evader P_2 at the first time T_{end} at which $x(t)$ enters the capture set $\mathcal{X}_{\text{end}}$. In case $x(t)$ never enters $\mathcal{X}_{\text{end}}$, the capture time T_{end} is $+\infty$. The pursuer P_1 wants to minimize an integral cost with variable horizon

$$J := \int_0^{T_{\text{end}}} g(t, x(t), u(t), d(t))dt + q(T_{\text{end}}, x(T_{\text{end}})),$$

whereas the evader P_2 wants to maximize it.

Example 18.1 (Canonical pursuit-evasion). Consider a pursuer P_1 that moves in the plane with maximum velocity v_1 and an evader P_2 that also moves in the plane, but with maximum velocity $v_2 < v_1$. The dynamics of this game are given by

$$\begin{cases} \dot{x}_1 = u, & x_1 \in \mathbb{R}^2, \ \|u\| \leq v_1, \\ \dot{x}_2 = d, & x_2 \in \mathbb{R}^2, \ \|d\| \leq v_2. \end{cases}$$

The pursuit is over when $x_1 = x_2$ and therefore the capture set is given by

$$\mathcal{X}_{\text{end}} := \{(x_1, x_2) \in \mathbb{R}^4 : x_1 = x_2\}.$$

We assume that the pursuer wants to minimize the capture time

$$J := \int_0^{T_{\text{end}}} dt,$$

whereas the evader P_2 wants to maximize it.

Note. See Exercise 18.1. ▷ p. 222

The Hamilton-Jacobi-Bellman-Isaac equation (18.1) for this game is given by

$$-\frac{\partial V(t, x_1, x_2)}{\partial t} = \min_{u \in \mathcal{U}} \sup_{d \in \mathcal{D}} \left(1 + \frac{\partial V(t, x_1, x_2)}{\partial x_1} u + \frac{\partial V(t, x_1, x_2)}{\partial x_2} d\right) \tag{18.10a}$$

$$= \max_{d \in \mathcal{D}} \inf_{u \in \mathcal{U}} \left(1 + \frac{\partial V(t, x_1, x_2)}{\partial x_1} u + \frac{\partial V(t, x_1, x_2)}{\partial x_2} d\right) \tag{18.10b}$$

$$= 1 - v_1 \left\| \frac{\partial V(t, x_1, x_2)}{\partial x_1} \right\| + v_2 \left\| \frac{\partial V(t, x_1, x_2)}{\partial x_2} \right\|,$$

$$\forall x \in \mathbb{R}^n, \quad t \geq 0 \tag{18.10c}$$

with boundary condition

$$V(t, x_1, x_2) = 0, \quad \forall t \geq 0, \quad x_1 = x_2 \in \mathbb{R}^2.$$

Intuitively, the time to capture should be a linear function of the initial distance between pursuer and evader, so the solution to this PDE should be of the form

$$V(t, x_1, x_2) = \alpha\|x_1 - x_2\| = \alpha((x_1 - x_2)'(x_1 - x_2))^{\frac{1}{2}},$$

for some appropriate constant α (to be determined). This function trivially satisfies the boundary condition and since

$$\frac{\partial V(t, x_1, x_2)}{\partial x_1} = \frac{\alpha(x_1 - x_2)}{\|x_1 - x_2\|}, \qquad \frac{\partial V(t, x_1, x_2)}{\partial x_2} = \frac{\alpha(x_2 - x_1)}{\|x_1 - x_2\|},$$

when we substitute V into the Hamilton-Jacobi-Bellman-Isaac equation, we obtain

$$0 = 1 - \alpha v_1 + \alpha v_2 \Leftrightarrow \alpha = \frac{1}{v_1 - v_2} > 0,$$

from which we conclude that the value of this game is given by

$$V(0, x_1(0), x_2(0)) = \frac{\|x_1(0) - x_2(0)\|}{v_1 - v_2}.$$

To compute the saddle-point state-feedback policies, we need to find the values u^* and d^* that achieve the outer minimum in (18.10a) and the outer maximum in (18.10b), respectively, which turn out to be

$$u^*(t) = -v_1 \frac{\frac{\partial V(t,x_1,x_2)}{\partial x_1}}{\left\|\frac{\partial V(t,x_1,x_2)}{\partial x_1}\right\|} = \frac{v_1(x_2 - x_1)}{\|x_1 - x_2\|}, \quad d^*(t) = v_2 \frac{\frac{\partial V(t,x_1,x_2)}{\partial x_2}}{\left\|\frac{\partial V(t,x_1,x_2)}{\partial x_2}\right\|} = \frac{v_2(x_2 - x_1)}{\|x_1 - x_2\|}.$$

□

18.5 Practice Exercise

18.1. Verify that for a given vector $c \in \mathbb{R}^m$,

$$\max_{x \in \mathbb{R}^m : \|x\| \le v} c'x = v\|c\|$$

with the optimal at $x^* := \frac{v}{\|c\|}c$ and

$$\min_{x \in \mathbb{R}^m : \|x\| \le v} c'x = -v\|c\|$$

with the optimal at $x^* := -\frac{v}{\|c\|}c$.

REFERENCES

[1] T. Başar and G. J. Olsder. *Dynamic Noncooperative Game Theory*. Number 23 in Classics in Applied Mathematics. SIAM, Philadelphia, 2nd edition, 1999.

[2] M. G. Crandall and P.-L. Lions. Viscosity solutions of Hamilton-Jacobi equations. *Trans. of the Amer. Mathematical Soc.*, 277(1):1–42, 1983.

[3] D. Fudenberg and D. K. Levine. *The Theory of Learning in Games*. MIT Press, Cambridge, MA, 1998.

[4] M. Grant, S. Boyd, and Y. Ye. *CVX: Matlab Software for Disciplined Convex Programming*. Stanford University, Palo Alto, CA, June 2008. Available at http://www.stanford.edu/~boyd/cvx/.

[5] T. H. Kjeldsen. John von Neumann's Conception of the Minimax Theorem: a journey through different mathematical contexts. *Archive for History of Exact Sciences*, 56(1):39–68, 2001.

[6] U. Madhow. *Fundamentals of Digital Communication*. Cambridge University Press, Cambridge, UK, 2008.

[7] J. R. Marden and A. Wierman. Distributed Welfare Games. *Operations Research*, 61(1):155–168, Jan. 2013.

[8] D. Monderer and L. S. Shapley. Fictitious Play Property for Games with Identical Interests. *J. of Economic Theory*, 68:258–265, 1996a.

[9] D. Monderer and L. S. Shapley. Potential Games. *Games and Economic Behavior*, 14:124–143, 1996b.

[10] J. Nash. Non-Cooperative Games. *Annals of Mathematics*, Second Series, 54:286–295, 1951.

[11] J. von Neumann and O. Morgenstern. *Theory of Games and Economic Behavior*. Princeton University Press, 1944.

[12] J. Robinson. An iterative method of solving a game. *Annals of Mathematics*, Second Series, 54:296–301, 1951.

[13] R. W. Rosenthal. A Class of Games Possessing Pure-Strategy Nash Equilibria. *Int. J. of Game Theory*, 2:65–67, 1973.

[14] L. S. Shapley. Some Topics in Two-Person Games. In M. Dresher, L. S. Shapley, and A. W. Tucker, editors, *Advances in Game Theory*, pages 1–29. Princeton University Press, 1964.

[15] J. von Neumann. Zur theorie der gesellschaftsspiele. *Mathematische Annalen*, 100:295–320, 1928.

INDEX